D0590803

HPL

70001061189 8

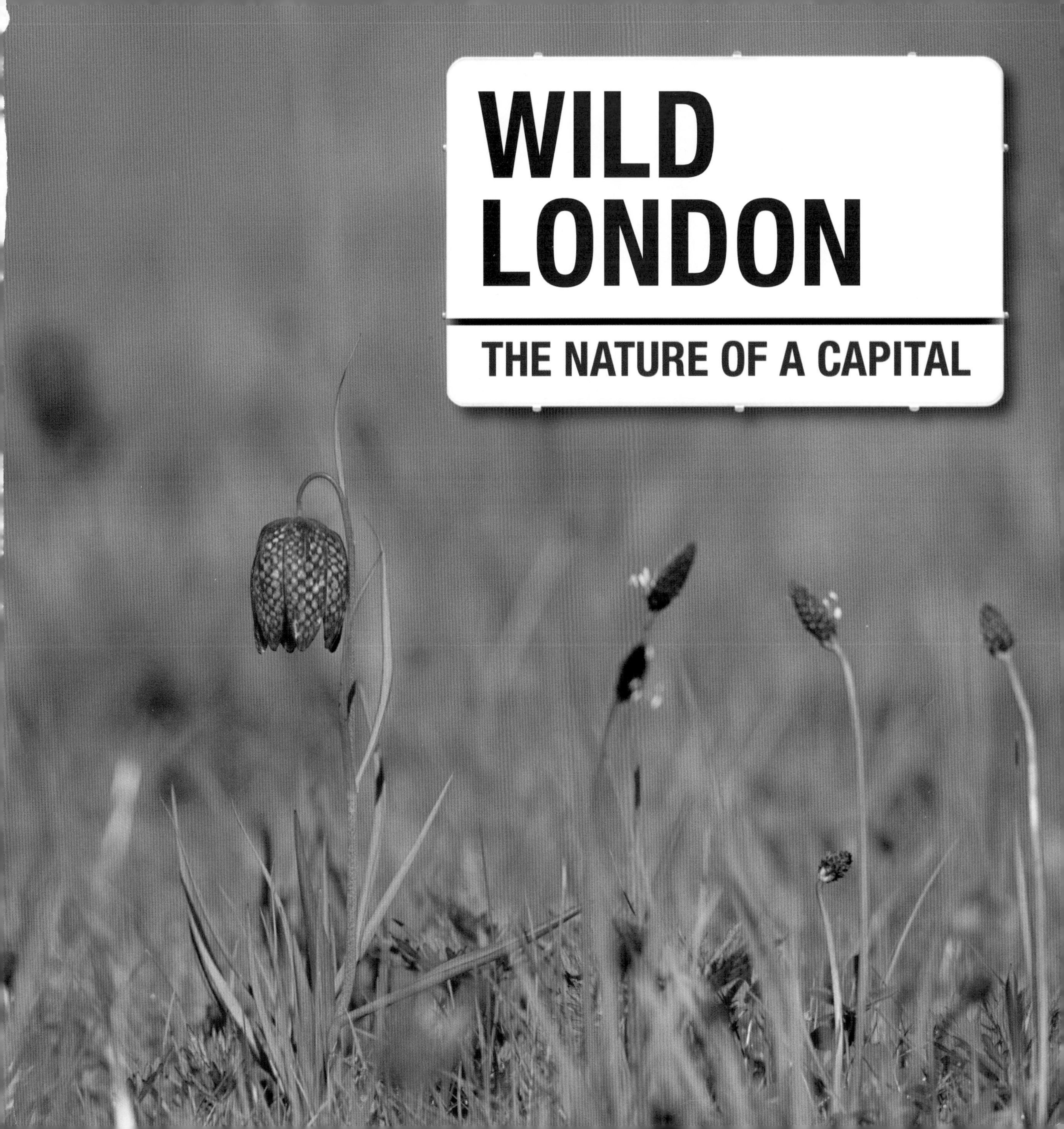
WILD
LONDON
THE NATURE OF A CAPITAL

For Ruth, Abbie, Sammy and for
my Mum and Dad who first inspired
and encouraged my interest in wildlife

www.tiger-books.co.uk
www.chevronpublishing.co.uk

Written and Photographed by Iain Green

Acknowledgements

While researching, photographing and writing this book I have drawn on the expertise and support of a huge number of people – many of whom have assisted me from its inception more than three years ago. To everyone who has helped me with this book, I extend a massive thank you.

In no particular order I would like to thank:

Sandi Bain and Andy Fisher for sharing their vast knowledge and offering extremely valuable guidance at the start of my research.

Stephanie Fudge and all the staff and volunteers at the fantastic London Wetland Centre. In particular, Martin Senior for his enthusiasm and unfaltering assistance; no request seemed too much trouble, even getting up early.

Colin Bartholomew, Sam Shippey and the whole RSPB team at Rainham, especially Nick Bruce-White for his tremendous assistance with my work on the Marshes.

Carlo Laurenzi, Catherine Harris, Tim Webb, Neil Ireland, Nicola Thompson and everyone at the London Wildlife Trust – an inspirational organisation looking after many of London's unique wild places. My special thanks to the staff at Camley Street – one of the locations that first fired my enthusiasm for London's wildlife.

Joanne Smith, Ben Dewhurst, Mick Wheatley, Tony Day, Rebeka Clark and all the passionate conservationists at the Trust for Urban Ecology – another superb organisation.

All the staff at 10 Downing Street for their fantastic assistance through the past few years and for allowing me very privileged access to the street's wildlife areas. And to Leo Blair for letting me share in his passion for the garden's mini-beasts.

Sarah Foraud and Trevor Williams for sharing their foxes and knowledge with me.

Chris Gittings for introducing me to the wonders of flounder and Deptford Creek – one of the most unique wild sites in London.

Christopher Raeburn and everyone at Phoenix Garden – a real wild treasure.

Angus Beveridge, David Mulholland, Clive Jones and Rhys Owen-Roberts at BAA for their assistance and allowing me special access to their conservation sites.

My old friend Nick Morgan and everyone at the Woodland Trust/Hainault Forest.

Dave Morrison and the Battersea Power Station team for sharing their black redstarts and peregrines with me.

Sophie Oliver and Alison Shaw at the Zoological Society of London's Marine & Freshwater Conservation Programme, for a tip that turned into one of my most rewarding days of photography.

Su Gough and John Marchant at the British Trust for Ornithology for so much useful information.

Also to Tony Drakeford, Sue Randall, Lynne Martin, Warwick Reynolds, Pete Massini at English Nature, Synthia Griffin at Tate Modern, Catherine Behan at London Underground, Simon Keith at the Whale & Dolphin Conservation Society and Des McKenzie, bird recorder for the London Natural History Society.

My thanks to The Royal Parks for permission to photograph in The Regent's Park, Hyde Park, Kensington Gardens and Richmond Park. Within the Royal Parks I would like to specifically thank Theresa Short, Nick Biddle, Denise Davis, Tom Jarvis and Jason Dudley-Malick.

Regent's Park rangers, Dave Johnson and Tony Duckett, deserve extra special thanks – without their support, this book would be poorer and I would never have discovered what Marylebone Road looked like at 4 am.

Thanks also to Keith French, Richard O'Mahoney, Cindy Blaney and the Corporation of London for greatly assisting me and allowing me to photograph in Epping Forest, Hampstead Heath and Highgate Woods.

I am extremely grateful to The Natural History Museum, in particular Gwyneth Campling, for allowing me to photograph in the wildlife garden and for providing information about fossil discoveries within London. My special thanks to Caroline Ware for kindly showing me 'her' garden.

And behind the scenes I would like to thank Paula Rodney for reading my manuscript under challenging circumstances. NHPA – my superb agents. Actpix in Wales for their care and expertise in producing the scans for this book. Richard Squibb and all at Vine House. Colin Woodman for enthusiastically turning my manuscript and photographs into a gorgeous book. Sally Forsyth, Eddie Creek and everyone at Tiger Books – especially Robert Forsyth and Michael Bird for believing in me and wholeheartedly supporting this lengthy project.

Finally to Ruth, my wife, for her constant support, unbounded encouragement and faith in me, even when my working hours were a little unsocial. And for first introducing me to the amazing wild spaces that make London so special.

And to everyone I met on route – thank you.

First published in Great Britain in 2005 by Tiger Books (UK)
an imprint of Chevron Publishing Limited

Friars Gate Farm
Mardens Hill
Crowborough
East Sussex
TN6 1XH England

Telephone: (44) (0) 1892 610490
tiger.books@chevronpublishing.co.uk
www.tiger-books.co.uk

ISBN 0-9543115-1-5

Project Editor - Robert Forsyth

Cover and book design by Colin Woodman Design

Printed in Singapore

Contents

Introduction

On a chilly pre-dawn summer morning last year, I found myself perched on the top of a church roof in central London. With me was Dave Johnson, who had agreed to me joining him on his daily bird surveillance. But this was no ordinary day. After months of waiting, the chicks of central London's first breeding pair of peregrines were due to take their first flight. Once threatened with extinction in the UK, peregrines were now nesting in London, even inside the 'Congestion Charge Zone'.

In the unusual quiet and serenity of the city, I watched the two chicks exercising their wings. An hour or so later, in the amber early morning light, the young female peregrine launched herself into the sky on her maiden flight.

It was an incredible privilege to observe and to take the first pictures of her first flight was even better. An experience never to be forgotten and one that, like all my studies, would not have been possible without the tremendous support and knowledge of local wildlife enthusiasts and conservationists.

I was first introduced to wildlife in London by my wife, a Londoner born and bred – we would watch kestrels hovering over Hampstead Heath and coots at Camley Street. Enthused by these early forays, particularly to Camley Street, which has become a personal favourite because of its unique situation, I began work on Wild London – a complete contrast to my studies of wild tigers in India, but just as exciting.

During the past three years I have explored London on foot, by boat, train, bus and car, visiting well over one hundred locations from gardens and roof tops to forest and wetlands. London may not seem the most obvious habitat for wildlife, but hopefully this book will show some of the city's wild wonders. Every day I spend exploring London, it becomes increasingly apparent what a fantastic diversity of wildlife the capital contains – orchids, voles, peregrines, salmon, stag beetles, seals and even dolphins. Far more than I could hope to cover in any depth in one book – thus my dilemma has been what to leave out, rather than what can be included. I hope that this book acts as a taster to the flora, fauna and wild habitats of the capital and encourages greater personal exploration and support for London's wild lands.

Wildlife watching in London, rather than further afield, has some real plus points. Firstly, most places in the capital are easy to get to – almost all the sites in this book can be reached by public transport. And secondly, because animals and birds are reasonably used to human presence, they can often be observed at a closer distance than in other parts of the country. The herons in Regent's Park are a prime example of how close you can get.

What is interpeted as London varies from person to person, often dependent upon where they live. The beauty of nature is that it knows nothing of our paper boundaries, so a precise explanation is academic. However, for simplicity my studies have been restricted to the area defined as Greater London – an area slightly smaller than the land circled by the M25 orbital motorway. A few wild places that straddle the border, such as Epping Forest and Rainham Marsh have been included.

Honesty and authenticity in wildlife photography are very important to me. All the photographs in this book have been taken within Greater London, apart from the hippo and hoopoe – for obvious reasons. All the animals and plants are wild and in their natural environment (the stag beetle was attending a London Wildlife Trust press launch).

As the regeneration and redevelopment of London continues at a fast pace, particularly in readiness for the 2012 Olympics, it is important to retain a firm hold on the importance of the city's green and brown habitats for wildlife and for people. Not only does London have a rich biodiversity heritage to protect, but people need open spaces and access to nature.

Iain Green
July 2005

Foreword

by Andy Fisher,
Head of Wildlife Crime Unit,
Metropolitan Police Service,
New Scotland Yard

London's wildlife is one of its less well-known treasures. The many visitors to the capital, and many Londoners themselves, see little of the wildlife that they share the city with, yet even amongst the hustle and bustle of 21st century London, wildlife is all around. The wildlife of London is surprisingly varied. Some species, like grey squirrels, are seen often, while others, like badgers and muntjac are more secretive. The black redstart is a London "speciality", and South London is a stronghold of the rare stag beetle. Today, Londoners can even see the spectacle of a peregrine falcon hunting the skies above Westminster itself. This is a genuine endangered species with the same international status as the tiger and the giant panda but which lives here in the heart of our capital.

This book is important because it lifts the lid on London's wildlife and some of its wilder locations and, through Iain Green's photographs and easy to read text, shows how nature can thrive, even in a city with a population approaching eight million. The pages that follow will take you from Rainham Marshes in the east to Richmond Park in the west, and even into the centre of Westminster. Wild London shows how much wildlife we have to value – sometimes in the most unlikely places.

To many people wildlife adds to the quality of our daily lives, but it needs our help. I hope that this book will inspire more of us to take an interest in London's natural heritage, and its protection, because if we value it, we will protect it, and this is a responsibility that we all share.

Nature of the Capital

Once the last visitors have left and the cemetery gates are locked, two fox cubs emerge from below ground to enjoy the warm spring evening. Obscured by lush vegetation, a fallen obelisk and headless statue become the cubs' playground. Awaiting their parents' return with food, the young foxes enjoy the peace and security of their surroundings – a wild oasis in a largely urban landscape. However, this is not neglected land forgotten by humans; it is the result of a positive conservation initiative by the local authority and neighbouring community. In varying degrees, humans have shaped every scrap of London. And although there is no virgin land, there is much wilderness. Nature works quickly and, when allowed, will swiftly colonise any space that is briefly left untidied. Narrow nesting ledges on tower blocks, reserves reclaimed from derelict industrial land or the uncut meadow in a city cemetery – this is Wild London.

Historically rich, culturally diverse and architecturally loaded, London is also wildlife wealthy. The city may not seem the most likely location for nature, but amazingly two-thirds of London's land area is green space or water.

The green space is roughly split into thirds – private gardens, parks/sports fields and other wildlife habitats.London's wild lands explore the extremes, from the expansive National Nature Reserves of Ruislip Woods and Richmond Park to railway linesides and innovative living roofs. Even in places devoid of greenery, some of the city's most exciting animals have made their home. Contrary to popular opinion, London's wildlife is not just limited to pigeons, starlings, buddleia and rats – the city's biodiversity is monumental.

"London's wild lands explore the extremes..."

Wren fledglings

It will be no surprise that the city pigeon is the most abundant bird in London. However, it is only one of 300 species that have been recorded in the capital. At the opposite end of the scale to the pigeon, ignoring accidental visitors, the rarest of London's birds must be the two bittern who have over-wintered in Barnes for the past few years. Other notable species include the peregrine, which has recently started breeding in the city and the black redstart, for which London is the UK stronghold.

According to a two-year study of London's birds by the British Trust for Ornithology (BTO), the blackbird is the most widespread species, found in almost all the sites surveyed. Even in the West End, blackbirds thrive in the peace of Phoenix Gardens, just off Charing Cross Road. A small community space, only a third of an acre, Phoenix hosts breeding pairs of blackbirds, wrens and blue tits. In late spring, visitors to the gardens can enjoy good views of parent wrens and tits feeding their young. The gardens are a remarkable place for wildlife in the hub of London and attract many other bird species, including great tits, whose London population has doubled over the past decade, as well as greenfinches, kestrels and even a garden warbler. Most surprising of all are the hundred or so frogs that somehow arrived in Phoenix Gardens and now flourish in this urban oasis.

Sadly, it is not all good news for London's wildlife – two of the capital's most famous birds have experienced serious drops in their numbers in recent years. The city's starling population has reduced by almost 20 per cent over the past decade, though this is slower than its decline nationwide. In contrast, despite sparrow numbers being stable nationally, the London population has suffered a 70 per cent decrease. The BTO study has shown that sparrows are generally absent from areas of ordered trees and short grass and are much more likely to be found in areas where understorey vegetation has been allowed to grow. The removal of shrubbery and undergrowth, for tidiness or safety, may be one of the reasons for the once abundant sparrow's demise.

Construction, a changing climate and land management practices, are all altering the future for London's birds. Over the past decade, the sparrow and starling have been two species on the losing side. However, the recent winners have included a diverse selection of birds, such as great tits, herons, blackcaps, peregrines and parakeets. Herons and rose-ringed parakeets have become two of London's most prominent bird species and blackcaps are now seen year-round in the capital. With continued development and warmer weather, London is set to see many more changes over the coming years, both arrivals and departures.

More than 1,500 species of flowering plants occur in Greater London, including the exquisite but scarce fly and bee orchids, the much-loved rosebay willowherb and the ubiquitous buddleia. In the 19th century, the rosebay willowherb was regarded as a rare species, yet it is now an abundant flower of varied terrain – from unpromising roadsides and building sites to verdant riverbanks and railway embankments. In a recent popularity survey, rosebay willowherb was voted as the capital's emblem flower. Growing in similar places to the willowherb, the buddleia softens the brick and concrete landscape. The buddleia's form, vibrant colour and honeyed smelling flowers are appreciated by Londoners and its nectar rich flowers are enjoyed by hordes of butterflies, bees and moths.

Juvenile blackbird

While some of the city's insects may be less than welcome visitors to our homes, gardens and woodlands in south London are a UK stronghold for the globally-threatened stag beetle. Britain's largest terrestrial beetle, the male can grow to about 7 cm or (2.5 in). The stag beetle is so named because the male's large jaws look like a stag's antlers. And it is the males who are most likely to be encountered, on warm summer evenings from May to August when they fly looking for a mate. Eggs are laid in the ground near rotting dead wood and the larvae can take up to seven years to reach maturity. With such a long larval stage, they are very prone to disturbance. This is one of the reasons they are such a threatened species. Wimbledon Common, Richmond Park and Dulwich Wood are some of the best places to find these impressive beetles.

Male stag beetle

Misrepresented in *The Wind in the Willows,* 'Ratty' was, in fact, a water vole. Observed briefly they may appear similar, but the brown rat and water vole are very different in their appearance, habitat needs and status. Unlike rats, water voles have small, concealed ears, a blunt rounded nose and their shorter tail is hairy.

However, the irony of the confused identity is that of London's mammals, the brown rat is probably the most abundant and the water vole is one of the most threatened. The loss of natural riverbank habitat is a key reason for the vole's decline, which is sadly credited as the UK's fastest declining mammal species. London is home to some important water vole populations – Rainham Marsh, Barnes Wetlands Centre and Fray's Farm, along the River Colne, are some of the best places to find them – look for the field signs.

The most obvious indications of occupied water vole habitat are ditches or riverbanks peppered with holes. The vole's burrows are long and, both for safety and accessibility, have many entrances, above and below water. Another visible field sign is the neat latrines, where water voles routinely deposit their droppings. At Barnes there is a latrine under a bridge, which the resident vole uses daily. Quiet observation a few metres from the bridge reveals one of these beautiful creatures walking along a well-worn path. Nearby, voles can sometimes be heard crunching on shoots of waterside plants, frustratingly close, but perfectly concealed. A distinctive '*plop*' is often the only clue to the vole's presence, as it drops into the water for safety. However, if you are lucky, you might glimpse one of these endearing shaggy rodents swimming across a ditch or large pond from burrow to food source. The most patient observer might be rewarded with the perfect 'storybook' image of 'Ratty' sitting on the bank chewing on some aquatic vegetation.

"A distinctive 'plop' is often the only clue..."

In contrast, the abundant brown rat is to be found in most London locations, from the House of Commons to the city's sewers. The easiest place to watch them is in some of the central London parks, where they can often be seen feeding on discarded food scraps on the edge of flower beds or shrubberies.

Including the rat and water vole, about 35 species of mammal occur in London – from deer to dolphins, shrews to seals. About a third of the species are bats and a further third rodents. Also thriving in the city are fantastic examples of fungi and fish as well as reptiles and amphibians.

On any day of the year, even within reach of a city worker's lunch break, there are numerous wild spectacles waiting to be discovered – some are dramatic, others subtle and require a little patience. Perhaps the simplest and most impressive is watching the wild cityscape change through the seasons. Observing the cycle of life, as migrant birds arrive and depart, vegetation grows or recedes and insects, fungi and flowers emerge with the seasons.

Two acres of accessible wilderness in the heart of the city, Camley Street Natural Park is only a few minutes walk from King's Cross station, or a pleasant canal-side stroll from Islington or Camden. Though small, the internationally acclaimed reserve is one of the finest examples of wild London. Explored through the seasons, life on the reserve is rarely quiet.

A thin strip of land alongside Regent's Canal, Camley Street reserve was originally a coal depot once owned by Samuel Plimsoll, inventor of the Plimsoll Line. The depot was perfectly located to move coal between barges on the canal and the surrounding railway lands. As the use of coal and the canal changed, the site became derelict and overgrown by buddleia. In the early 1980s, there were plans to build a coach park on the land, but the idea was soon dropped in light of strong local opposition. Following a community campaign, the site was earmarked as a nature reserve and in 1983 work began on landscaping the area and constructing a large pond.

Opened in 1985, Camley Street is today one of the London Wildlife Trust's flagship

Winter
Completely frozen over, the ice on Camley Street's pond gives clues to the wanderings of a resident coot.

reserves and a vital green sanctuary for people living in a highly urbanised part of London.

The reserve has become a haven for wildlife. Reed warblers and chiffchaffs frequent the reed beds. Newts, frogs and toads breed in the pond and in the watery margins, marsh marigold and yellow flag iris are dominant. The woodland of hazel, silver birch and hawthorn provide cover for wrens or hammering posts for greater spotted woodpeckers, while the meadow offers a perfect environment for many butterflies including the stunning common blue. Feeding stations around the visitor centre attract garden birds and the adjacent canal brings herons, swans and ducks to the park.

However, despite its immense natural and community value, the reserve has lived an uncertain life. On a brief walk through the park, across the reed bed boardwalk and into the woodland, you could be forgiven for forgetting where you are. A tiny green island amid a sea of construction, the Camley Street wilderness lies in the centre of a vast building site – some say the largest in Europe.

Spring
By April, reeds are starting to grow and leaves on trees disguise the truth about the reserve's location.

Summer
Although only five minutes' walk from King's Cross Station, reed warblers have bred in the reedbeds at Camley Street.

Until plans were adjusted, Camley Street was to be buried under the new Eurostar terminal at St Pancras – they now sit alongside each other. Ironically, during the building of the St Pancras terminal, many of the construction workers took their lunch break in the tranquillity of the nature reserve.

Spurred by the construction of the Channel Tunnel Rail Link and the Eurostar terminal, King's Cross is undergoing a massive regeneration programme. Houses, office blocks, shops and cafés will transform the brownfield site that was once industrial and railway land.

It is testament to the tenacity of city wildlife and the co-operation between developer and conservationist that Camley Street does not seem to have been too adversely affected by the construction traffic, dust and noise pollution.

However, the park continues to tread a fine line and the building works are planned to continue on a large scale for many years. There are plans for a high-rise office block at the end of the reserve and a wide footbridge over the canal, cutting through the park. Regeneration of the area will also bring increased traffic and less obvious pollutants such as street lighting.

While the surrounding 'wasteland' is changing beyond all recognition, only time will tell what the future holds for Camley Street Natural Park. However, with the London Wildlife Trust fighting to protect and conserve it and many other of London's special places, the reserve stands a decent chance.

Autumn
Gentle green and golden foliage gives visual relief from a landscape dominated by bricks and concrete.

Greenfield or Brownfield – Coloured Vision

In any thriving city, there is continual conflict between leaving space for wildlife and the use of land for development. The richness of London's biodiversity is thanks to the efforts of the many individuals and organisations who over the decades have fought to protect the city's wild habitats. Camley Street, Oxleas Wood, Gunnersbury Triangle and Rainham Marsh are prime examples of what can be achieved by people power.

To avoid building on green fields and the associated negative publicity, there is an emphasis on regenerating areas of 'wasteland' or brownfield sites as they are better known. But wrongly labelled as 'wasteland', there is a growing awareness that the derelict brownfield sites are rich in biodiversity. London's unused urban spaces offer a welcome wild chaos to what could be a rather too tidy city. However, 'space is money' and viewed by planners as second-class habitat, brownfields are the prime sites for redevelopment. Species that live in these areas, such as the nationally endangered black redstart, are under increasing pressure as their habitats are cleared.

In terms of biodiversity, the beauty and importance of brownfields is that they are often closed sites and may have lain undisturbed for years. Colonised naturally by wildlife and growing unmanaged by humans, brownfield sites are unique wild spaces and characteristic of the city life. It is too simplistic to differentiate potential development sites for construction by their colour or aesthetics. Brownfield sites perfectly complement the city's managed green spaces and many deserve similar protection.

Wetlands and the Urban Chill

Harassed into the smouldering plum sky by carrion crows, a ghostly flock of almost two hundred black-tailed godwits move between Aveley and Wennington Marshes. Almost silhouettes, the distinctive dark outline on their underwing is just visible in the half-light. As the afternoon sun recedes behind Rainham landfill site and tower blocks on the southern edge of the Thames, the birds nervously settle in a tight group in the shallows of a large pool. The uniform, pale-grey plumage of the sleek godwits instantly distinguishes them from the many lapwings, which also rest in the marshes. The chill winter air encourages a calm to descend upon the birds. Through the cold months the city's precious urban wetlands are the temporary feeding grounds for many thousands of wintering waders.

Greylag geese

Dreaming of a white Christmas, tradition portrays the December festive season as mid-winter. However, it is in January and February that the true wintry weather takes hold. Snow rarely falls in London until well after the New Year. With milder winters and temperatures a few degrees higher than less urban areas, snow is an uncommon sight in the heart of the capital. In a changing climate, snow may become an even rarer sight in London.

But when it does snow, the face of the city changes dramatically – the distinctive features, blemishes and scars of human habitation are fleetingly obscured under a pristine white veil. Covered by a few inches of snow, the central parks are stunningly transformed; they become playgrounds for people, but a challenge for the wildlife that must continue to find food in their new surroundings. Hampered in their grazing by the rare layer of snow and the frozen lake, greylag geese approach any human who may be willing to part with some food.

Some birds, such as grey herons, become incredibly courageous in harsh weather. To the amazement of people living in rural areas, the normally elusive riverside fisher joins the duck and pigeon feeding parties that await park visitors. Sometimes the juvenile herons come as close as to within two or three metres of human passers by.

Juvenile grey heron

Fieldfare

Encouraged by our own desire to 'hibernate', the reduced daylight hours and inclement weather narrow our view of the natural world in winter. Many plant and animal species have entered a dormant state and our attention is often focused on the birds and mammals entering city gardens. The seasonal bonanza on the bird table and mammals sometimes literally on the doorstep can offer wonderful and unusually close wildlife viewing. Yet further afield from our homes, there is a surprising wealth of wild activity in winter and in some of the most unlikely locations. In London parks, large mammals can be observed easily and elsewhere the vegetative decay provides a greater chance of viewing the city's smaller mammals. Many bird species are now at their most obvious, in large communal roosts, wintering feeding flocks or nationally important gatherings of wildfowl.

The chill brings shutdown for many plants and insects, hibernation for reptiles and amphibians and long periods of sleep for bats, but there is a bustle of bird and mammal life.

Except for hedgehogs and dormice, terrestrial mammals do not enter complete hibernation, rather choosing to sleep through cold spells and emerging to forage on milder days, when they must work harder than normal to find food. From secretive mice to foxes and deer, London's mammals adjust their routines, often relying more than normal on humans for food and shelter.

"a bustle of bird and mammal life..."

In Sydenham Hill Wood, a tiny wood mouse survives the worst of the winter by feeding on stores of acorns gathered in autumn and by reducing its energy consumption. Emerging from a complex burrow on a mild February day, the mouse can look forward to warmer weather and the imminent arrival of spring shoots and bulbs to feast on.

Migrating southwards, birds such as fieldfares and redwings have come to the capital for a less severe climate than their breeding territory. These two wintering thrush species are often seen in large mixed flocks on sports fields and parkland or devouring the autumnal berry bonanza in hedgerows.

For some of the finest wildlife watching in winter, head for the water. Wildfowl, waders and other birds gather in large numbers on the city's plentiful water-bodies. These are some of the capital's finest wild habitats. From Barnes Wetland in the west, along the River Thames to Rainham Marsh in the east and north to the extensive Lee Valley – in winter, life in London's wetlands is booming.

"the marshes are a precious wilderness area with a wonderful future..."

Collectively, the contiguous marshes of Aveley, Wennington and Rainham form the Inner Thames Marshes Site of Special Scientific Interest (SSSI). Situated on the eastern edge of the capital, amid a heavily industrialised area, the marshes are one of London's largest undeveloped habitats.

In 1906, the War Department purchased the property for use as a rifle range, a function that continued for much of the century. Restricted access to the land over this period meant that the wildlife and medieval marshland system were left relatively undisturbed (Wennington Marsh has been identified as an intact remnant of late medieval marsh with features dating from the 14th century). With the end of Ministry of Defence activities, the site was put up for sale and began to be abused by trail-biking, falconry and other destructive pursuits. Neglected and earmarked for commercial development, conservationists began a lengthy campaign to protect the wetlands. Thankfully, through the dedication of local people and environmental groups, the marshes were saved. In July 2000, the Royal Society for the Protection of Birds (RSPB) bought 871 acres of the land, 77 per cent

of the total site, creating Rainham Marshes Nature Reserve, its first reserve in Greater London. The future of the remaining 23 per cent was secured in 2002, when the London Borough of Havering pledged to protect the final portion of Rainham Marsh.

The Inner Thames Marshes are immensely valuable because they support such a wide variety of wildlife in nationally important numbers – and not just birds. The three marshes host one of the highest densities of water voles in the UK. Designated as a 'Key Site' for water voles, the marshes are so significant because of the assured long-term protection they offer these vulnerable creatures. Also found on the reserve are 69 species of nationally notable invertebrates and a dozen scarce or uncommon wetland plants.

However, the wetlands are at their best in winter when huge concentrations of waders and ducks are to be found. Most obvious are the teal and lapwing whose populations often number thousands. Other ducks such as wigeon, shoveler and gadwall and waders such as dunlin, golden plover and redshank can occur in many hundreds.

Though parts of the site are still affected by development, such as construction of the channel tunnel rail link that runs along the northern rim, the marshes are a precious wilderness area with a wonderful future. The outlook may be even brighter, for there are plans that once the adjoining landfill site ceases operation within the next decade, it will be incorporated, along with the marshes, into a proposed conservation park. Encompassing approximately 1,600 acres, this would almost double the area protected for wildlife.

"...a thousand-strong flock of lapwings"

Given a casual glance, the Inner Thames Marshes can appear bleak, devoid of life and unworthy of so much fuss. However, upon closer inspection it is clear that these windswept wet grasslands are packed with birds. Instantly obvious, when a thousand-strong flock of lapwings suddenly lifts into the air, spooked by a large lorry as it rumbles along the adjoining construction site road. Birds take flight from along every marshland margin and even the smallest grassy islands. The huge lapwing flocks are choreographed in their movements, flying one way, then wheeling in the opposite direction. These avian 'clouds' change colour as they turn, from dark plumage to light underside, then back again with another roll. As the lapwings gain height, slowly the aerial acrobats lose formation and small parties break away. With the threat gone, it is safe to settle once again in the marsh. A calm spreads across the wetlands, encouraged by the soothing sounds of the lapwing's distinctive '*pee-a-wit*' call.

Some of the lapwings are resident and will stay and nest in the drier grasslands, but most will disperse to farmland, where spring-sown crops provide good nesting sites.

'A leap with a waiver in it' is the literal translation of the Old English 'hleapwince', from which the lapwing gets its name. It aptly describes the male's captivating territorial display – rising into the air, gaining height on slowly beating wings before suddenly dropping towards the ground, rolling and twisting in a large arc.

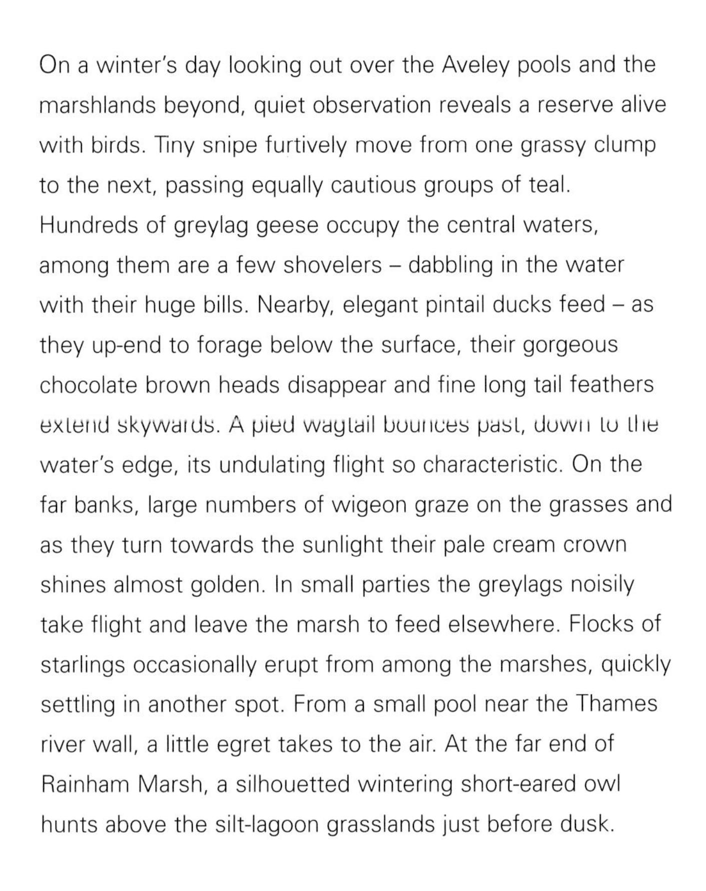

Little egret

On a winter's day looking out over the Aveley pools and the marshlands beyond, quiet observation reveals a reserve alive with birds. Tiny snipe furtively move from one grassy clump to the next, passing equally cautious groups of teal. Hundreds of greylag geese occupy the central waters, among them are a few shovelers – dabbling in the water with their huge bills. Nearby, elegant pintail ducks feed – as they up-end to forage below the surface, their gorgeous chocolate brown heads disappear and fine long tail feathers extend skywards. A pied wagtail bounces past, down to the water's edge, its undulating flight so characteristic. On the far banks, large numbers of wigeon graze on the grasses and as they turn towards the sunlight their pale cream crown shines almost golden. In small parties the greylags noisily take flight and leave the marsh to feed elsewhere. Flocks of starlings occasionally erupt from among the marshes, quickly settling in another spot. From a small pool near the Thames river wall, a little egret takes to the air. At the far end of Rainham Marsh, a silhouetted wintering short-eared owl hunts above the silt-lagoon grasslands just before dusk.

"flocks of starlings erupt from among the marshes..."

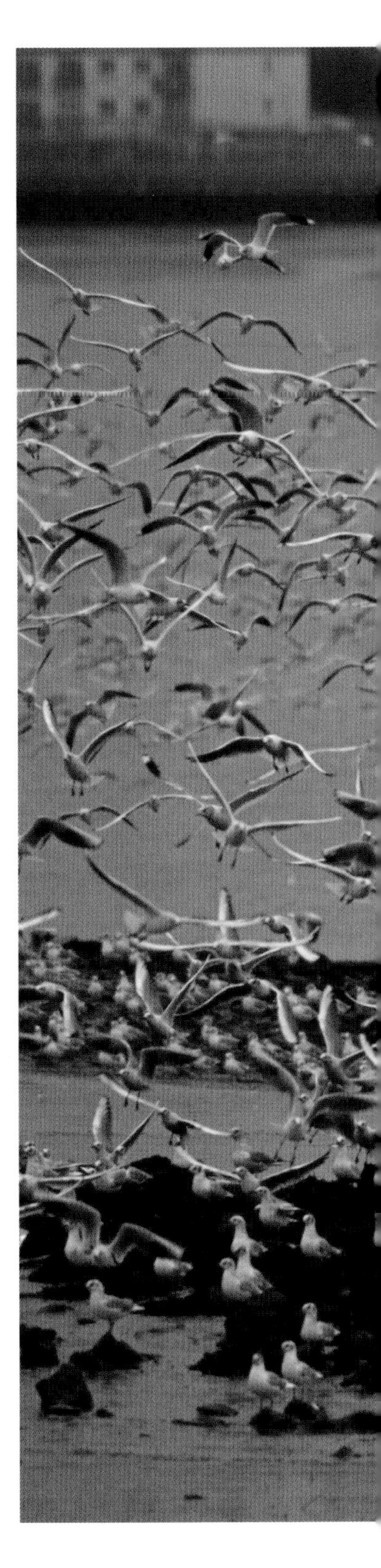

Massing in the marshes, numbers of teal on the RSPB reserve can peak at more than 3,000 in January. The smallest of all British ducks, teal are very fast flyers and tend to move in tight irregular groups. They are secretive birds, preferring to feed in small pools where their size means they can easily hide.

Protected from human disturbance by the marshes, the mudflats of Aveley Bay are a safe resting place. Teal numbers dramatically increase in winter, when migrant birds arriving from Northern Europe join the small UK breeding population.

Flocking at a truly urban wetland, thousands of gulls erupt from the foot of a deep-sea-diver. Alongside Rainham marshes to the south-west lies the Havering River Path, a prime and characterful location for birds on the Thames Estuary. From Coldharbour Point to wild land adjoining the lonely Tilda Rice Factory, there are a variety of attractions, both natural and man-made. The footpath passes the imposing, but pungent, landfill site, the strange steel 'diver' sculpture and some disused concrete barges.

Enticed by the freely available food supply, gulls feed on the refuse tip in tens of thousands. Primarily black-headed gulls, which can number 15,000 in mid-winter, they forage on the tip, then flock on the Thames foreshore or the marshes for a much-needed bath and drink. Herring, common, lesser black-backed and greater black-backed gulls also occur in significant numbers and this area is probably the best place in the capital to see the more unusual yellow-legged gull.

A few hundred metres further north along the Thames edge are some Second World War barges marooned on the mudflats. Originally used in the D-Day landings, the massive concrete structures were deposited in 1953 to form part of a flood defence system. Firmly embedded and surrounded by vegetation, these war veterans are now the favoured habitat for wintering water pipits.

In the murky waters below Tower Bridge swims a seal. Over the past few hours, the common or 'harbour' seal, has leisurely swum the two miles up-river from Rotherhithe, feeding along the way. Despite the image that the seal's name might conjure up, this is definitely not a common occurrence in Central London.

Slowly working its way along the southern edge of the Thames, the seal seems quite at home. Underwater for a few minutes at a time, it regularly surfaces, occasionally with a fish and once with a plastic bag. Sometimes London's gulls harass the seal as it bobs in the water, probably trying to steal some freshly caught fish. Inquisitive of the boats and buildings along the edge, it comes very close to the foreshore. Then, when almost at the bridge, the seal turns and with a sudden rush swims at full speed a short distance down-river. Slowing down, it resumes its relaxed pace, heading back to Rotherhithe. For the moment, this is the seal's patch.

A few days later, amid a light snowfall, the seal swims up-river as far as the Millennium Foot Bridge, crosses over to the north bank and spends an hour feeding and observing the goings on by the Tower of London.

Amazingly, these are not isolated incidents; over the past year, there have been a number of sightings of common seals and also some of its larger cousin, the grey seal. The cleaner Thames offers good fishing and at low tide the foreshore provides many discreet beaches where the seal can haul itself out. All this one requires now is some like-minded companions!

However, seals are not the only marine mammal visitors to the Thames. Over the past few years, there have been sightings of other species. Both bottle-nosed and common dolphins have recently been observed along the Thames, up as far as Blackfriars. During Christmas 2004, three harbour porpoises were spotted swimming under Tower Bridge.

Drawn by the significant improvement in the water quality of the Thames, it is wonderful that marine mammals are starting to return to London. What cetacean species will be the next to visit? In 1949, a narwhal was recorded just outside the London reaches of the Thames. The narwhal is a very distinctive species of whale, normally inhabiting the Arctic Circle – the male has a unique two metre long tusk.

The 'Thames Marine Mammal Sightings Survey' has been launched by the Zoological Society of London. The survey's aim is to gather information from anyone fortunate enough to spot dolphins, porpoises and seals in the Thames.

For survey forms and to report any sightings contact, marineandfreshwater@zsl.org

---

Today, the Thames has the seal of approval; it is one of the cleanest metropolitan rivers in the world. Yet it has not always been attractive to aquatic life. In recent centuries, the river water had become heavily polluted, incredibly smelly and almost devoid of fish. Ever since people first came to the Thames thousands of years ago, the river and its wildlife have been increasingly shaped and pressured by the human presence.

"... one of the cleanest metropolitan rivers in the world."

Before human intervention, the river was much wider than it is today; several channels meandered along with mudflats in between. Many small tributaries joined the Thames, which was flanked by marshland and forest. The first people settled in the area about 6,000 years ago and they began to clear the forest for agricultural land. Remains of ancient forests that grew 5,000 years ago can still be seen on the foreshore at low tide, in places like Erith and Chelsea. In the 1st century AD, the Romans established 'Londinium' where the present day city now stands and ever since, the metropolis has expanded. The growth of the city is intrinsically linked to the story of the Thames. The mighty river is not only the dominant feature of London, it is the reason the city exists at all.

Before the Thames reaches Greater London at Hampton, it has flowed for more than one hundred miles from its source, a spring in Trewsbury Mead, Gloucestershire. Emerging from the Cotswolds, the junior Thames eventually reaches Cricklade, site of the famed fritillary meadow. On its voyage eastwards, the river runs through Oxford, Reading, Henley and Windsor and just outside the M25 orbital motorway, the Thames passes Runnymede, site of the signing of the Magna Carta. After winding through the cluster of reservoirs on the outskirts of London, the Thames reaches Bushy Park and Hampton Court, one of Henry VIII's favoured palaces.

A short distance beyond Hampton, at Teddington Lock, the Thames becomes tidal, although it is still freshwater. Under normal conditions, the freshwater zone continues to London Bridge and from there onwards it is brackish – where fresh and salty water mix. Gradually the river increases in salinity, but is not seawater until Southend, well outside the London reaches.

Through the medieval period, the Thames was tamed and shaped. Much of the river was embanked and the surrounding marshland began to be drained. Slowly, the river became narrower and its edges increasingly built-up.

Population and building pressures over the past few centuries have brought an even worse fate to a number of London's other smaller rivers. Of the many tributaries that once flowed through London, some like the Lee and Wandle still wind their way to the Thames, others have been lost forever. North London's Fleet River now runs underground as part of the capital's sewerage system. Once busy with sailing barges, it was covered over from Holborn to Blackfriars in the 1700s and today, stripped of any dignity, carries the city's waste. The Effra that once flowed through Brixton was diverted during the building of the Albert Embankment and is now an enclosed sewer.

With the growth of London, increasing pressures were placed on the river. The rise in riverside industries such as slaughterhouses, tanneries and other factories saw escalating amounts of industrial pollutants enter the Thames. An expanding population brought more sewerage, much of which ended up being poured straight into the river. Years of being abused as a highway for industry and waste brought the Thames to its knees.

In the mid 19th century, four outbreaks of cholera killed 35,000 people. Until this time the link between water and cholera had not been realised. Then in 1858, an excessively hot summer resulted in 'The Great Stink'. Pleasure trips on the river ceased and the Houses of Parliament had to drape perfumed curtains at its windows to mask the offensive smell.

The solution was the construction of the Victoria, Albert and Chelsea embankments to house main sewers, which would carry London's sewerage away from the city centre. Opened in 1865, 1,300 miles of sewers took the waste to outflows on the Thames at Beckton and Crossness. It was many more years before any cleaning of the sewerage took place. Despite the success of the sewerage system, the growth in London's population continued to cause a decline in the water quality. Thermal pollution from power stations, polluted tributaries and factory discharge meant that by the 1960s, the Thames was almost biologically dead in parts. The only surviving fish were eels, a species that can live in both fresh and salt water.

In the 1970s, changes to sewerage treatment and the prevention of industrial pollution enabled the river to stage a remarkable recovery. Though it may appear murky, the brown colouration is, in fact, only silt stirred up by the tidal currents. Thanks to tight controls on what goes into the river and hard work by bodies such as the Environment Agency and Port of London Authority, the water quality in the Thames now supports an incredible diversity of flora and fauna. Well over a hundred species of fish now feed in its waters, including flounder, bass and smelt, a species known to need good water quality. Salmon even migrate up it to breed. Three hundred and fifty invertebrates and 300 plant species are also found on the tidal Thames.

LONDON'S ANCIENT NATURAL HISTORY

Under the capital's streets and buildings lie the remains of a very different landscape – where unexpected wild creatures roamed. Like the whole of Britain, London was shaped by the ice age, or rather a series of 'ice ages' separated by warmer times. Each period of warmth brought a distinct natural heritage to what we now know as London and clues to the city's ancient residents are continually being discovered.

During the construction of Trafalgar Square in the 1830s, builders unearthed the remains of bones from crocodiles, elephants as well as a hippopotamus jaw. About 120,000 years ago, this area may have resembled the African plains of today.

Not far away, the Georgians and Victorians found evidence of giant bison, wolves, sabre-toothed cats and woolly mammoths in the London clay – animals of a different time and environment. Reindeer fossils have been discovered in South Kensington and under Battersea Power Station were the remains of a woolly rhinoceros.

Recently, the neck bone of an auroch (a giant cow) was discovered on a building site in Knightsbridge. Identified by experts at the Natural History Museum, the bone is 170,000 years old. As parts of the city are redeveloped or transport tunnels dug, there will be many more finds – further evidence to piece together London's ancient wild history.

"the vibrant wetlands at Barnes bring a breath of fresh air to the city..."

On World Wetlands Day, 2nd February 2002, the London Wetland Centre was designated a Site of Special Scientific Interest (SSSI), only six years after work began to re-wild the site. Just a few miles from Westminster, four uniform concrete reservoirs have been transformed into one of London's most extraordinary and precious havens for wildlife. Against a tradition of filling wild spaces with concrete developments, the vibrant wetlands at Barnes bring a breath of fresh air to the city. Formerly known as Barn Elms, the reservoirs became redundant upon completion of the Thames Water Ring Main, a 50-mile water tunnel under the capital. Already an important site for wintering wildfowl, the owners, Thames Water, wished for a sympathetic use for the land. In a unique partnership, they leased the site to the Wildfowl and Wetlands Trust, who undertook a massive project to landscape and convert the reservoirs into varied wetland habitat. The remarkable transformation was funded through many sources, including the building of houses on a small portion of the old site

In May 2000, the London Wetland Centre was opened by Sir David Attenborough. An impressive visitor centre and six viewing hides allow close observation of the 105 acres of lakes, pools, grazing marsh and reed beds, without disturbance to the wildlife. Though still in its infancy as a nature reserve, the Wetland Centre at Barnes is already host to an impressive array of resident and visiting species, including almost 400 moth and butterfly species and 180 bird species.

On the banks of the Thames, the varied wetland habitats are a magnet for wintering wildfowl. One major draw for the birds is that being so close to human development the water will be one of the last areas to freeze in harsh winters. The list of visiting ducks is as impressive as the birds themselves – teal, pintail, gadwall, wigeon and shoveler. The peak winter count for ducks is about 1,500 birds. Large numbers of shoveler reside on the main water, with their distinctively enlarged bill. Like whirling dervishes, pairs or small groups of the shovelers spin round in circles or figure of eight patterns, sieving food from the water. Towards the end of winter, male shovelers can be seen posturing at rival ducks. This dominance display involves raising their heads and bobbing the bill up and down.

Other ducks like mallard, tufted and pochard are found all year as they breed on the reserve. For the time being, ruddy ducks are abundant, but native to America and introduced by humans to the UK, they are now the focus of a controversial nationwide cull. The wetlands also attract more unusual wildfowl, including goldeneye, garganey, common scoter, smew and even those, like shelduck, that prefer coastal waters.

Mute swans are plentiful and a solitary bewick swan, normally only a winter visitor so far south, has taken up a lonely residence – it is generally seen feeding along the margins of the reeds. Greylag and Canada geese are a common sight grazing on the flooded marshland.

Water birds aside, the wetlands are a fantastic lure for birds of prey and owls. The combination of an exceptional city habitat and the fact that the site is well watched by birders has given the centre an impressive bird list. Long-eared and short-eared owl are observed most winters and kestrel, sparrowhawk, peregrine and hobby are commonly seen. More unusual raptor visitors include, merlin, hen and marsh harrier, goshawk and even osprey.

Female mallard

Reed bunting

Bitterns are the most famous avian visitors to the Barnes wetlands. They first arrived in January 2002 and at least two birds have returned every winter since. Related to herons, bitterns are extremely elusive birds that live in dense reedbeds. Loss of habitat across Europe has caused the population of this species to plummet. Now very rare in the UK, there are perhaps 20 breeding pairs and in winter, numbers may swell to almost a hundred. That such an endangered bird chooses to return regularly to the wetland, clearly states the immense value of this city centre habitat.

The wintering bitterns generally stay in Barnes for a few months, whipping up a frenzy of birdwatchers and other visitors eager to catch a glimpse of them as they move silently through the reeds. As if the bitterns' streaked chestnut brown plumage is not good enough camouflage, they may remain motionless for long periods to further blend in with the reeds. Often the clearest sighting is as they fly from one area to another, swooping across the water, almost owl-like with their short compact wings.

However, they are birds of habit and each of the bitterns prefers to stand in particular spots and fly at certain times. After weeks of observation, their movements may be guessed with a fair degree of accuracy.

Though they are mostly silent birds, in the breeding season males make a booming sound that can be heard a mile away. Booming bitterns in Barnes is many a birdwatcher's dream, but highly unlikely as the reed bed area is thought to be too small to support a breeding pair. However, if more areas of reed bed were created nearby, then you never know. For local conservationists, it really would be something to make noise about.

"...swooping across the water, almost owl-like with their short compact wings"

Bittern

Snipe

Only a stone's throw from central London, sitting in the Wetland Centre's two-storey Wildside hide, visitors have a superb panoramic view across the wetlands. Looking north is the deep reservoir lagoon, a favoured spot for wintering ducks such as shoveler and teal. Remarkable congregations of cormorants rest on the small islands or wooden posts and there is the chance of an occasional sparrowhawk in the trees on the reserve boundary. To the east is the grazing marsh where large parties of coot and wigeon feed on the semi-aquatic vegetation. Herons stand at regular intervals along the water's edge and in particularly cold conditions a pair of snipe probe more exposed parts of the marsh near the hide. South are the reedbeds and channels, not only home to the famous bitterns, but also equally secretive water rails. Above the water, flitting from stem to stem are reed buntings and other small passerines. The view west looks over some smaller pools, the winter domain for many resident birds, mute swans, moorhens, mallards and great-crested grebes.

Teal

On a cold January afternoon, one of the resident foxes at the London Wetland Centre escapes the wind on a sheltered bank. Blending into the background, a number of visitors pass by without noticing it. Eventually, yawning and stretching, it is time to move on – a truly urban fox, it seems little concerned by human presence.

"...a truly urban fox"

The undisturbed habitat and ample foraging opportunities make the Barnes reserve a good place to live. Though they may occasionally take wildfowl, the foxes are a natural and important part of the ecosystem. The diet of a city fox is, perhaps not surprisingly, a varied one and quite balanced. Ducks aside, on the menu are insects, small mammals, berries, discarded processed food and vegetative waste.

"...the park feels like a remote Scottish moor."

Under darkened skies, Richmond Park is unusually devoid of human visitors and, bar the occasional parakeet calling, silence settles in the grasslands. A group of red deer stags rest in the long grass as snow starts to fall. A kestrel aborts its hunting foray and perches in an old oak tree on the stump of a long-since fallen branch. Resigned to rest as the light flurry becomes blizzard-like, the hovering hunter surveys the inclement conditions. For the few remaining hours before dusk, the park feels very bleak, far removed from urban London and more like a remote Scottish moor.

At dawn on a radiant winter morning, a chilly Richmond Park lies beneath a thin white sheet. The Pen Ponds have remained largely unfrozen, a blessing for the many gulls, geese and ducks that rely on them. Smaller pools have completely frozen over, however. Snow adorns every tuft of grass, but in the daytime warmth it will not last long – the melt has already started.

Richmond's fallow deer have moved to a sheltered area of woodland. Only a light sprinkling of snow covers the bracken, but the soil is still frozen. Fallow stags seek out a pile of recently cut tree prunings to browse on. In the warming mid-morning sun, they bathe and casually spar.

"a flock of brilliant-green rose-ringed parakeets swoop past..."

Lining the boundaries of Lewisham Crematorium is a row of leafless poplars. Each evening, a section of these tall trees plays host to an unlikely avian extravaganza. Almost accurate enough to set your watch by, about ten minutes before sunset, as the light begins to fall, a single screeching '*kee-aw kee-aw*' can be heard in the distance. Within seconds, a flock of brilliant-green rose-ringed parakeets swoop past and land in the willow trees at the centre of the crematorium. Before long, flocks of 20 to 30 of the vocal birds arrive at speed from every direction. They are easily identified in the air by their distinctive silhouette and swift, steady flight. During the next thirty minutes, between one and two thousand birds alight in the willows, which shake with the commotion of the parakeets. Some fly down to briefly drink from a small stream, a few individuals seek out last-minute snacks in nearby trees. The rush-hour gathering in the willows is not the parakeet's final destination, for suddenly they take to the air and fly to the nearby poplars, settling in the very tops of the trees. Down below, commuters hurry along the main road to their own brick roosts, with barely an upward glance. Only an occasional 'look at the parrots' can be heard.

Descended from caged pet birds that escaped in the 1960s, the wild population is now so established that the parakeet is classified as a British species. Native to India and other parts of Asia, they are accustomed to both hot summers and cold winters and deal with the British climate quite happily. Recent studies by the British Trust for Ornithology have shown that parakeet numbers are increasing rapidly, though they are not extending their range at a similar speed. Approximately 10,000 roost within the M25/Greater London area.

Before the sun rises, the parakeets will leave their night-time haunts to return to favoured feeding grounds across London from Beckenham Place Park to Hampstead Heath.

Silhouetted against a dusk sky, the parakeets are a spectacular sight. Though darkness has descended, there is no hush as the birds continue their shrill, noisy chatter. This is London's loudest if not largest roost.

However, any natural phenomenon continually evolves and what may be visible one day might have disappeared the next. Despite the capital's growing parakeet population, the winter of 2004/05 saw an unexpected and dramatic drop in the numbers roosting at the crematorium. It is thought that the colony split in two, one half having found a new roost site elsewhere. Perhaps the colony became too big or maybe the birds moved because of human disturbance. The poplars

lining the crematorium still contain a noisy, if reduced, population of parakeets and maybe in a few years time they will once again shake under the weight of thousands of birds.

Parakeets may be the most colourful and noisy of the colonial roosters, but pied wagtails are certainly London's most elegant. A gracefulness in appearance and movement beautifully illustrated by their Italian name of 'ballerina'.

Like festive decorations, dainty pied wagtails adorn the tree branches. Often missed by the hurrying crowds of chilly late-night Christmas shoppers, large flocks of the wagtails roost in built-up areas during the coldest nights. In urban and suburban parts of Greater London, the 'dancers' gather together for safety and maybe warmth, often assembling in a solitary tree.

Starlings were once famous for their city centre dusk displays, however, these are not as common as they were. Like sparrows, starling numbers have significantly decreased in urban areas and subsequently roost sizes have shrunk.

"...dainty pied wagtails adorn the tree branches like festive decorations"

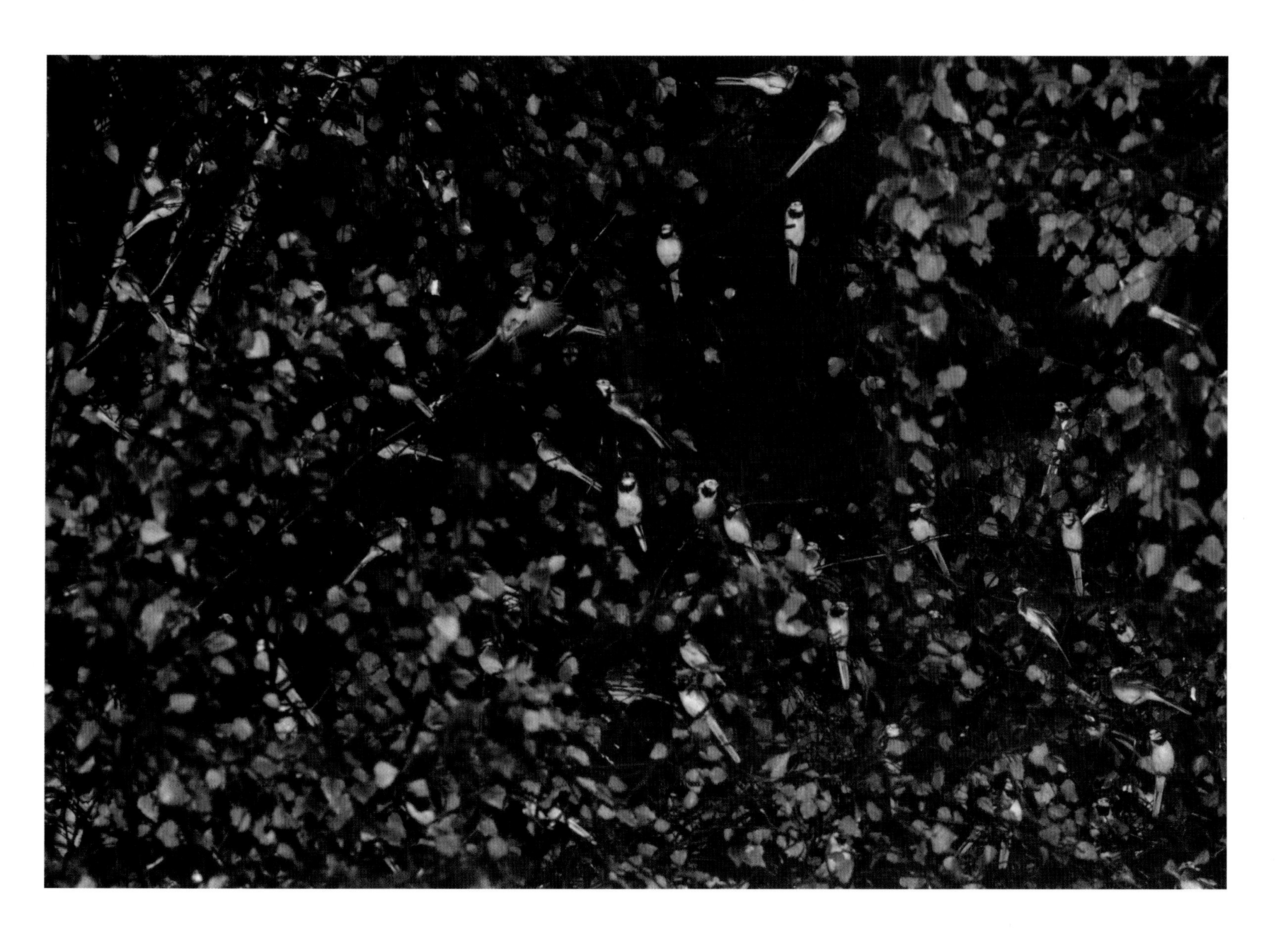

Not an obvious hotspot for wildlife – only five minutes' walk from Tottenham Hale underground station – the Walthamstow Reservoirs offer superb birdwatching opportunities and are home to two of London's most impressive bird colonies.

On a windy winter day, the exposed reservoirs can feel like one of the coldest places on earth; the raised paths running on top of the banks receive the full force of any cold winds. Even more windswept are the few islands protruding from the centre of the reservoirs. Two islands in Reservoir 5 are home to a massive colony of cormorants, which roost and nest in the trees – in past years, roosting numbers have reached 500. The birds fly up the Lee Valley in late afternoon, from their daytime fishing haunts along the Thames. However, in recent years many of them have become semi-resident, feeding by day on the good fish stocks to be found in the ten adjoining reservoirs.

Relative newcomers to London, cormorants first bred in the capital as recently as 1987. A common sight today, they began nesting at Walthamstow in 1990 and the colony now numbers in excess of two hundred nests and is continuing to grow. Receiving little protection from the elements, the island trees may not seem a great nest site. However, for this traditionally coastal species used to nesting on cliffs and rocky ledges, the birds must feel quite at home.

The neighbouring Reservoir 3 is home to a significant nesting colony of grey herons, one of the largest heronries in the UK. Since 1916, the herons have bred here and now almost a hundred nests are occupied. The herons begin nesting in mid-winter and can already be incubating eggs in February. It has been suggested that the dramatic increase in cormorant numbers might adversely affect the herons through a reduction in fish and nesting availability. As yet there is no apparent conflict, which may be because they use different islands and that the cormorants lay their eggs in spring, much later than the herons.

Dropping down from the trees to water level, the surface of the reservoirs also holds much of interest during winter. Nationally important numbers of tufted duck gather in rafts, together with other species such as pochard and the more elusive goldeneye. Great-crested grebes are abundant in the cold months, feeding on the plentiful fish stocks. Rarities like divers occur occasionally – individual birds that have strayed inland, far from their usual maritime habitat.

Valentine's Day may have just passed, but the Welsh-Harp drizzle veils romance and the signs that spring is coming. Looking across Brent Reservoir – also known as Welsh Harp because of its shape – in the distance, the new Wembley Stadium is close to completion. Wembley's arches offer a beautiful backdrop to a pair of courting mute swans. Pairing for life, the swans continually reaffirm their pair bonds through display.

On a cold, wet day in mid-February, the reservoir in Hendon may not instantly appeal, but out on the water one of the most passionate wild spectacles is taking place. A pair of great crested grebes is engrossed in an elegant courtship display. Prior to the start of the breeding season, both sexes grow distinctive double crests of dark plumage on their heads, chestnut ear tufts and a neck ruff. Due to an insatiable demand for the elegant plumes from the fashion industry, the great crested grebe almost became extinct a century ago. In 1860, less than 100 birds survived in the UK, however, last-minute protection reversed the decline and the UK population is now a healthy one.

Nationally important numbers breed on the Welsh Harp reservoir and now is the time for their refined courtship. Beautiful to watch, pairs face each other and perform a sequence of displays, involving head shaking, feather fluffing and diving. Often described as a 'penguin dance' the grebes swim towards each other and by paddling rapidly, rise out of the water, breast to breast and bill to bill. Treading water and with a vegetative offering in their bill, one of the pair presents the plant to its mate. This courtship will continue until the pair eventually mate.

Along the edge of the reservoir, blackthorn bushes already have significant sprays of flowers and nearby, blackbirds and great tits are singing as if it were already spring.

Birds may have begun breeding, but the weather can still hold surprises. Late February and early March can see some of the coldest winter weather; the chance of significant snowfalls and frozen ponds or ditches is still very real. Varying widely from year to year, at some point during late February and March, winter will pass. The beauty and drama of spring will soon unfurl.

Great crested grebe

Capital Growth

Enhanced by the soft morning light, the attraction of a solitary snakeshead fritillary is immediate. The stunning chequered flower is mostly deep pink, almost purple, but one in every few plants produces delicious creamy white blooms, with a hint of green. Sadly, this plant of damp grasslands and river meadows is now rare and only found in certain locations across London – Darland Lake Nature Reserve, the London Wetland Centre or at Camley Street Natural Park, where a few grow. Like spring, the snakeshead spectacular arrives suddenly and its beauty passes at a swift pace – the fritillaries flower for only a brief spell in April. Fantastic early sources of nectar for insects such as this garden bumblebee, by May Day most will be past their prime. Following closely behind the fritillaries are seas of glorious bluebells, the playful days of young fox cubs and the first flypast of screeching swifts.

Swathes of delicate blackthorn blossom announce the imminent arrival of spring. The flowers first appear in February, well before the leaves emerge. By March, a great ivory veil cloaks the hedgerows and thickets. Flowering so early in the year, this snowy covering may not be the only one the blackthorn receives.

"...a cold spring was known as a blackthorn winter"

Often growing alongside hawthorn, these two native species are firmly established in traditional folklore and have flowers of similar appearance. However, that is where the likeness ends for they beautifully symbolise the journey from the cold tail of winter to the warmth of early summer. Because the blackthorn flowers appear when easterly winds could still bring bad weather, a cold spring was known as a 'blackthorn winter'.

In contrast, not until May does the hawthorn display its sweet-scented white flowers. Also known as the May tree, the clusters of blossom, each head spotted with pink anthers, herald the end of spring. There is much folklore surrounding the hawthorn, from pagan and medieval traditions relating to the arrival of summer to the belief that destruction of a hawthorn may bring peril. Offering nectar for insects, great nesting sites for birds and berries rich in vitamin C in winter, the destruction of the hawthorn hedgerows certainly does spell disaster for wildlife.

Some of the best sites in London to view the beautiful spring blossom are at Ten Acre Wood, Welsh Harp Reservoir, Selsdon Wood and The Chase Nature Reserve.

For perfect whites of another kind, the swathes of cow parsley and mass of horse chestnut blossom in Morden Hall Park are extremely impressive in early May. A National Trust property, Morden Hall is a superb sanctuary for wildlife and residents from the neighbouring area. Despite the park's urban surroundings, the last stop on the tube's Northern Line regularly attracts buzzards – a species traditionally associated with open country.

Soon the signs of spring are everywhere – shoots of new vegetation appear from below ground, buds emerge on twigs or branches and early flowers bloom. Golden yellow is the

colour of early spring. Daffodil, dandelion, marsh marigold, and lesser celandine bring the first warm hues to a landscape that has not yet fully woken. Following on from the rich yellows is a floral procession through suburban spring of intense colour.

The final proof that spring has arrived is the mass human exodus from inside to outside on the first warm day of the year. Cocooned in heated houses throughout winter, we emerge to be reminded of the forgotten wonders of a natural world. Greeted by a spring in full swing, this is the season of growth in the capital.

Bumblebees are some of the earliest insects to appear and not long afterwards the first butterflies fly by. Emerging from hibernation, peacock butterflies can be active from March onwards. Along the south bank of the Thames, opposite Canary Wharf, brown-tail moth caterpillars are emerging from communal silk webs, in which they have over-wintered. The airborne hairs of these caterpillars are an irritant and can cause discomfort to local people.

Many of the wetland birds that wintered in Barnes, Rainham Marsh and their other haunts will have left and will be well on the way back to their more northerly breeding grounds. But by late spring, the city's wildlife will be joined by thousands of summer migrants, from swifts and swallows to numerous species of warblers. Temporary residents, these warm weather visitors have come to London to breed.

Spring sees the best of the bird nesting season and perhaps surprisingly, London offers some of the most impressive and clearest viewing of communal and individual nest sites. Large nesting colonies of herons, tree sparrows, cormorants and sand and house martins are easily viewable. Nesting terns, grebes, coots, moorhens and kestrels can all be watched at reasonably close quarters without disturbance to the birds. And for the ultimate comfort, one of London's rarest nesting birds can be watched live on the internet via webcam.

But while enjoying the visual, do not forget to tune into the audio. In woodlands, woodpeckers begin drumming, announcing their ownership of woodland territory. Eight or ten blows to a tree trunk, delivered in a single second create the unmistakeable sound of a greater spotted woodpecker.

At daybreak on a spring morning, the dawn chorus is one of the most memorable and enchanting wildlife experiences to be enjoyed anywhere. Even in the heart of London, the mass of parks and gardens come alive with the symphony of bird song.

A season of growth, there is too much to absorb and too few days to spend doing so. Much will pass us by, but for the missed experiences of spring, there is always next year.

If modern London had a bird emblem, then the heron might be it. They are not rare, and they are also not common countrywide, but there are few places in the capital where today they cannot be found nearby. However, this has not always been the case. In his book *London's Birds* published in 1949, RSR Fitter stated; *'It can be asserted with a fair degree of certainty that neither the heron nor the spoonbill will ever again breed in London...'* Primarily fish-eating birds, the grey heron population had been hit hard in the early 1900s by lack of prey in the heavily polluted River Thames. Slowly disappearing from the capital, significant improvements in water quality have allowed the heron to survive, and indeed, prosper. Standing almost a metre high, grey herons have adapted with surprising success to life in the city, residing in the many parks, canals and rivers. Including the unlikely located heronry at Walthamstow Reservoir near Tottenham, there are more than 15 heronries in London. They even nest in the heart of the metropolis – less than a mile from Oxford Street.

The bizarre and highly visible heronry at Regent's Park, only a few metres from the busy roads around Baker Street, is the most accessible place to see them. In 1969, two decades after Fitter made his damning forecast, a pair of herons successfully raised young in the park. The following spring, ten pairs produced 16 young and, today, the heronry accommodates about 25 pairs. The nests are a mixture of natural, heron-made stacks of sticks and man-made woven willow saucers. It was once believed that heron nests had two holes in the base for the birds to dangle their long legs through!

One of the earliest bird species to breed, the park herons begin nesting in January. By spring, there are already chicks and visitors to the park are rewarded with the unusually clear view of parents feeding their hungry youngsters. As leaves clothe the trees, some of the nests become less visible, however their presence is easily betrayed by the noisy young and regular fly-pasts of an adult returning to or leaving the nest. By late spring, most of the young will have left the nest and joined the adult birds feeding along the water edge below. At this time of year, Regent's Park must have one of the highest densities of herons in the UK.

"...it was once believed that heron nests had two holes in the base for the birds to dangle their long legs through!"

The tawny owls of Regent's Park must hold the record for producing the earliest chicks of the year – though the herons breed early, the owls beat them wings down. Tawny owls normally lay their eggs in March, yet by late February, a pair in Regent's Park already have well-grown chicks. It is staggering to consider that the female must have laid the eggs in mid-December, almost three months in advance of tawny owls elsewhere. This is the third year that the pair have bred months ahead of the norm. It is likely that the cushion of warmer city-centre temperatures, milder winters and the resultant unseasonal abundance of active rodents stimulated the early laying.

In a hollowed tree trunk or nest box, the female incubated the eggs for a month before they hatched. The chicks remained in the nest site for a further few weeks before climbing into the tree canopy. Known as 'branchers' the young cannot fly at this age and are still reliant on their parents for food.

Peering through the canopy of a small wooded area in the park, the two chicks are experiencing one of the coldest days of the year. Already making short flight hops, they seek the most sheltered spot in the evergreen foliage. Keeping a discreet watch on the youngsters, the adults day-roost in two separate trees nearby.

By late-March, the young have taken their first proper flights and throughout April, are flying regularly. Having lost their downy feathers, they are sometimes difficult to tell apart from their parents.

The most common owl in Greater London, tawny owls are best known for their familiar hoot; though this is a common misconception. '*To-whit–to-whoo*' is actually two distinct calls – '*hoo-ho-ho-hoooo*' and a high-pitched '*ke-wick*'.

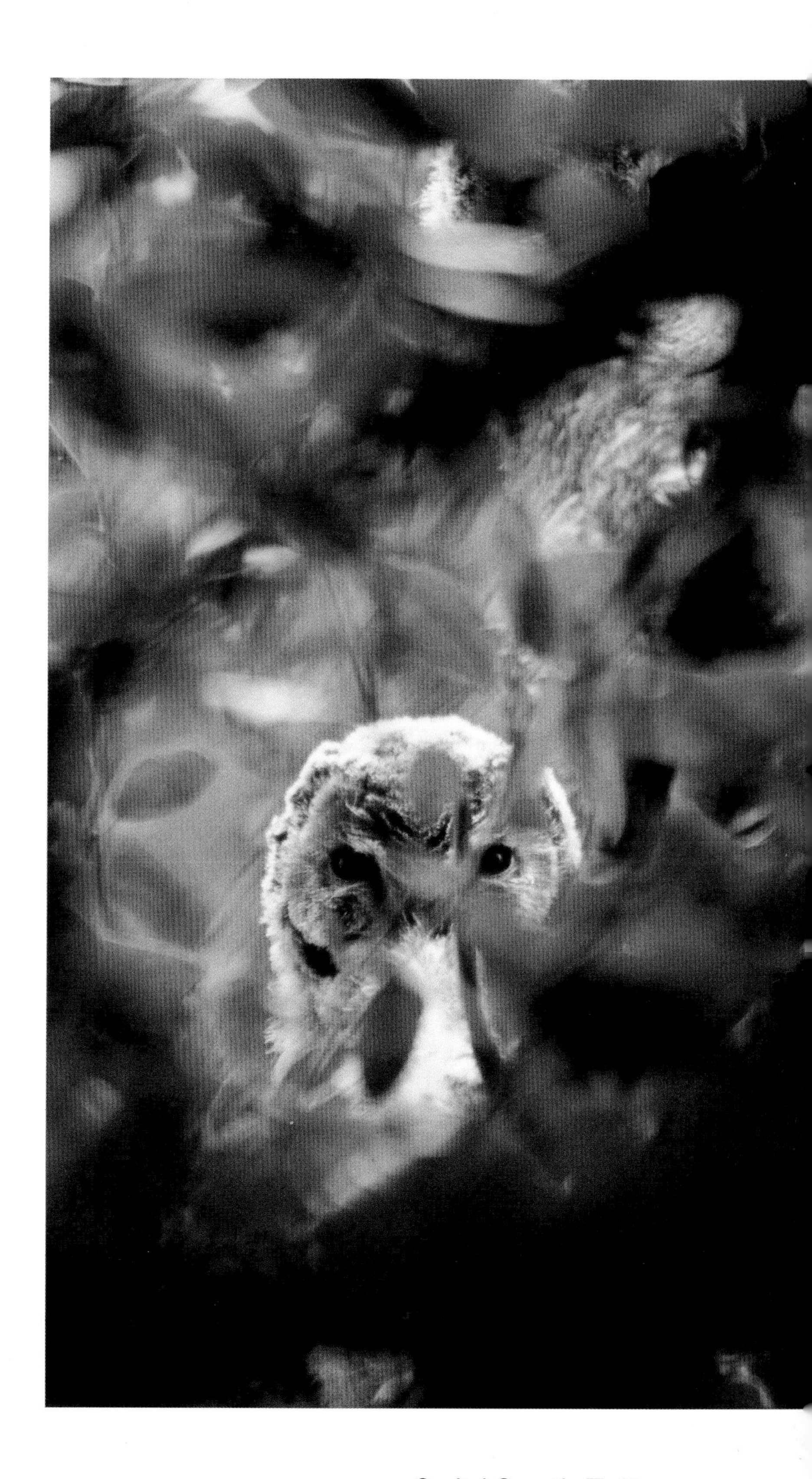

One of London's newest residents, rose-ringed parakeets are adapting well to life in the city. In many parts of London parakeets are regular visitors to bird tables and hoppers, feeding alongside more traditional garden birds.

Beckenham Place Park is a favoured breeding site for the parakeets, where they seek out old holes in trees in which to nest. In early spring, the parakeets are busy courting, which involves a lengthy display by the pair. Easily identified by the neck-ring, which the female lacks, the male slowly sidles along the branch to his mate. Repeatedly stretching his wings over the female, he gently preens the nape of her neck and eventually mates with her. However, the courtship is not over, the pair continue a complex bonding and kissing ritual. Sitting slightly apart, the male sways towards the female and they delicately link beaks. Separating for a while, he repeats this swaying and kissing a number of times before the pair eventually settle together side by side.

Frolicking frogs in garden ponds mirror courting birds in the trees above. By March, frogs have entered the peak of their spawning season. Common frogs are able to breed in colder weather than toads and often having hibernated within the pond, are the first of the city's amphibians to reach their breeding site. Each day, the pond's population swells and the frogs become livelier. The water begins to 'boil' with frenzied mating activity. Pairs swim around together, the male attached to the back of the larger female, his front limbs wrapped around her chest. Sometimes two males may struggle to take ownership of a single female. Though generally silent animals, this is the time of year to enjoy the croaking frog chorus.

In garden ponds, frogs casually tease their eager human observers as to when the first spawn will appear. Then suddenly one morning, a huge jellied mass of frogspawn floats in the water. Many pairs of frogs may choose the same spot to lay their eggs. Even though the spawn arrival is expected, it is still a surprise. An award of quality for the wildlife gardener.

Unlike frogs, toads only require an aquatic habitat in which to breed and may migrate some distance to return to a preferred pond. They have an amazing navigational sense and will cross many boundaries, natural or man-made, to reach their breeding water. Roads are one of the biggest killers of toads. In spring, the sad remains of squashed toads all too clearly mark these migrational highways. Thankfully, where the routes are well known, conservation measures, including warning signs, toad patrols and even road underpasses have helped to reduce the carnage.

Having negotiated the obstacles en route, the tired females now face danger of a very different kind. They may arrive at the water already carrying the smaller male, but if not, they will quickly be sought out by one, or many, males. Where there is a shortage of females, the bizarre phenomenon of toad-balls may occur – females are literally encased by a mass of clambering males. In their overpowering attempt to mate, the mob of males may unwittingly drown the female. But if the female is lucky enough to survive, she will lay a long string of eggs, which her mate simultaneously fertilises.

After about two weeks, the jelly surrounding both toad and frog spawn starts to disintegrate and the tadpoles swim free. They grow rapidly in size, and during the summer the new froglets and toadlets will emerge from the water to seek cover in thick vegetation

As well as the toad and common frog, London is home to the introduced and well-established marsh frog. Breeding later in spring, the marsh frog is a noisy species. The London Wetland Centre is a great place to appreciate its croaking cacophony.

All three species of newt occur in the capital, though the great crested and palmate newt are rare and difficult to observe. The smooth or common newt can be found in many sites across the city, perhaps the most unlikely place is directly underneath the flight path at Heathrow Airport.

"...enjoy the croaking frog chorus"

Terminal for wildlife? Heathrow Airport must surely be the one place where wildlife fears to tread. Yet as noisy, stark and developed as the urban world surrounding the runways may be, there are seven established conservation sites at Heathrow. They range from constructed wetlands, rivers and reedbeds, to wood and scrubland. And the irony is, that despite the inescapable aircraft roar some of these reserves are the least physically disturbed habitats in the capital. Restricted access at a number of the sites for safety and security reasons allows the birds, mammals, amphibians, reptiles and plants to live with little human interference although some of the habitats are carefully managed to ensure that species considered hazardous to aircraft are not encouraged. Airport authorities do not look kindly on large flocks of birds.

British Airports Authority (BAA), which owns Heathrow Airport, manages its conservation sites for maximum biodiversity. The company's environmental team has instigated detailed species studies and drawn up a thorough biodiversity strategy, which is gradually being implemented. According to this plan, BAA's aim is '...to ensure that the airport has an overall positive impact on local biodiversity.' However, it must be added that, globally, the air industry is far from wildlife-friendly and the growing demand for flights is a significant contributor to climate change.

Three hectares of phragmites reed beds at the Heathrow Constructed Wetland, provide great feeding and nesting sites for both reed warbler and reed bunting – a species in decline. The surrounding arable land provides valuable habitat for breeding skylarks – another species in national decline. The eastern side of the site is open to public access as a cycle track and green corridor.

"...a unique mix of man-made and wilderness environment"

On the eastern edge of Heathrow are the Causeway Nature Reserve and adjoining Camp 4 site – jointly a Site of Metropolitan Importance. The River Crane runs along the reserve's north-western boundary, bringing natural biodiversity to this isolated habitat. The 25-hectare site is principally two large reservoirs surrounded by mature willow woodland, grassland and reed beds. Apart from occasional access for organised visits, they are closed reserves, concealing a unique mix of man-made and wilderness environment.

When Heathrow Airport was built in the 1950s, the construction workers lived at Camp 4. Now half-covered by moss, much of the camp's concrete foundations still exist and some of the original street lamps still stand among the goat willow and white willow woodland. Being so close to the airport, it is impossible to hear anything in Camp 4 when aircraft pass overhead. Any wildlife living on this site must adapt to the oppressive noise and vibration of very low aircraft. However, detailed studies of the site's spider populations have revealed 118 species, including four that are nationally notable. Ninety-five species of beetle have also been recorded, including six that are nationally notable. The good invertebrate diversity is in part, due to the nutrient poor grasslands that grow on very thin soil over the hard standings.

Slightly further away from Heathrow than Camp 4, the Causeway Nature Reserve is marginally less noisy and holds a real wealth of wildlife. In spring, juvenile smooth newts absorb the warmth beneath specially placed squares of corrugated iron or black roofing felt. Similar in appearance to adults, which have now returned to their breeding ponds, the juveniles are not sexually mature and so will stay on land for about two years before heading to the water to breed. Carefully lifting the edge of one of the felt or iron covers might reveal three or four of the golden coloured amphibians and maybe even a toad. Marsh frogs are abundant here and they, in turn, support a large population of grass snakes. The reservoir holds decent numbers of waterfowl and five species of warbler breed in the area. Unusual bird visitors have included a black-necked grebe and a night heron, a vagrant from southern Europe.

Along the water's edge, oak, ash and willow are the dominant trees and away from the reservoir are stands of coppiced hazel. Clearings in the riparian woodland yield swathes of red dead-nettle, foxglove and red campion. Bee and common-spotted orchids will appear in drier areas later in the season. Growing over the site's old gravel entrance is a mosaic of tiny low-nutrient-tolerant plants, interspersed with daisies and moss.

It is clear that the seven established conservation sites are of great value to wildlife. However, two more sites have recently been added at Heathrow and one of these illustrates the conflict between wildlife and air travel. To satisfy increasing demand for flights, Heathrow is growing – construction of the new Terminal 5 is well underway. Destined to be destroyed by development of the approach road, an area of flood meadow of the River Colne has been translocated by BAA to a new protected site. However, less lucky was Perry Oaks Sewerage Farm, the main site for the new terminal. Until building began, the sewerage farm was a prime habitat for wintering waders and well known by birdwatchers.

Juvenile smooth newt

In the heart of the King's Cross building site, a pair of coots attend to their young at Camley Street Natural Park. Small pieces of vegetation from submerged plants are collected by the adults and passed to their chicks. Young coots are easily identified by their bright red bald heads and the ruff of yellow tinted downy feathers. Coots breed early in the year, from March onwards. The chicks are reliant on their parents for about only eight weeks after hatching, so the parents may lay a second, or even third clutch of eggs.

On the edge of Greenwich Ecology Park, the resident pair of mute swans have become local stars because of their prominent nest site. The female, or pen, does most of the incubation of the eggs, which will hatch after about 35 days.

Well adapted to life in London's water bodies, moorhens, like coots, can be incredibly bold nesters. Still under construction, this nest is within an arm's-length of a busy walkway.

"moorhens can be incredibly bold nesters..."

Singing for a female sedge warbler, a male launches into his varied chattering and melodious song. On a fine spring morning, the waterside hedgerows at the Barnes wetlands are lined with boasting birds. At regular intervals and in clear view, males boldly advertise for mates. April is probably the best time to see these normally elusive summer visitors; once the male has attracted a female, he will stop singing and descend into the thick vegetation. More than a dozen pairs nest in the reserve each year.

On early mornings in April, many of the capital's garden birds are singing at their finest. As its name suggests, the dawn chorus is at its best before most people are awake and though it can be enjoyed while still in bed, a pre-sunrise outing is highly recommended. In the rare quiet of the dawning city, the liquid and bubbling tune of the blackbird can be hypnotic; as the songster's luxurious sounds float through the air, they saturate the senses. Blackbirds are one of the first to break into tune, closely followed by robins; then wood-pigeons, dunnocks, wrens and great tits. The pleasure is absorbing the whole chorus, but also in listening to each different song, enjoying the '*teacher – teacher*' of a great tit and the '*coo*' of a wood-pigeon.

As beautiful as the chorus is, each song has a very serious purpose. That it is only the males that sing may be a surprise, but the song's function is to reinforce territory ownership and to advertise to females. Why particularly at dawn? There are a number of reasons. Obviously the world is a quieter one, so song is not drowned out. Also, because insects are inactive in the cold, the singing birds are not missing valuable feeding time. But there is another reason; the peak of the chorus coincides with the dawn egg-laying of females. As soon as she has laid, the female is at her most fertile and needs a male to fertilise the next egg. By singing at his best, the male songster keeps the female's attention firmly on him and sends out a clear message to other males that may potentially intrude. Once the early morning urgency has passed, the male can relax a little. Slowly, as the sun rises and the human population becomes active, the song subsides or is drowned out by the city din.

Sedge warbler

"the wetlands support a staggering 120 nesting pairs of reed warblers…"

Another summer visitor to London, the reed warbler's tune is more even and repetitive than that of the sedge warbler. It is a regular sound across the capital, as they breed in many locations. The wetlands at Barnes support a staggering 120 nesting pairs each year. Superb sightings of these secretive birds can also be had at the Greenwich Ecology Park, where well-located hides allow very close views.

Reed warbler

Duck Wood's bluebell display is one of the best in the capital, both vibrant and accessible. And the bluebell is the nation's favourite flower according to a survey by Plantlife, the UK plant conservation charity. There are few scenes as gorgeous and enriching as a sea of bluebells washing across a woodland floor. A carpet of vivid green leaves and stems support a profusion of purplish-blue flowers. Timed to perfection, the blooms erupt en masse in late April – before the fresh tree foliage has thickened and darkened the forest. The flowers perfume the air with a delicious, yet fresh fragrance. An early morning visit to Duck Wood, near Harold Hill, is the best time to enjoy the bluebell feast. While you are there, keep a look out for badgers, foxes, fallow deer or even mandarin duck, an exotic species that has become established here and in a few other London sites.

Ragged robin and breeding redshank are mid-spring arrivals at the London Wetland Centre. Patches of pink and white flowers of ragged robin thrive in the grasslands; surrounding them are swathes of glorious yellow buttercups.

Out on the water, the birdlife is just as rich; many species use the reserve as a brief refuelling stop on their spring migration. Wading birds such as curlew, ruff, godwit and sandpiper take advantage of the muddy feeding grounds at the reserve's wader scrape. Other waders like the little ringed plover, lapwing or this redshank stay to breed on the constructed islands dotted throughout the lake. Looking over the mudflats is an artificial cliff nest site, built to attract sand martins. The martins first used the man-made tunnels in 2004, when twelve pairs nested – the following year 25 pairs successfully raised young.

Attractive to common carder bees and a host of other insects, the subtle and slender flower buds of the yellow flag iris open to reveal brilliant golden beacons. The gorgeous blooms stand more than a metre high, flanked by bright green sword-like leaves. A flower of marshes, river margins and various wet habitats, the flag iris can be found in many wetland habitats across London, from late spring through to mid-summer.

An insect mimic, the elusive and exotic-looking bee orchid survives in a few chalk downland sites in London. The flowers attract male pollinators by mimicking the appearance of a female bee and even emitting female scents. Saltbox Hill, near Biggin Hill Airport, in the far south of London is probably the best place to search for them. Be prepared to look hard, for unlike the iris, they are small and do not wave flags.

"...mimicking the appearance of a female bee"

Common spotted orchid

Orchids of many types thrive at Saltbox Hill, one of the richest wildlife habitats within Greater London. Recently acquired by the London Wildlife Trust, Saltbox Hill SSSI (Site of Special Scientific Interest) is a small but impressive fragment of chalk downland – part of the North Downs. In late spring and summer, the grasslands positively buzz with life. Orchids, such as bee, fly, man, common spotted and pyramidal, grow well here, the latter two species are found in some abundance. The fields of orchids, cowslips and grasses are beautifully balanced by a mass of hawthorn blossom in the hedgerow, best at this time of year.

Prior to the Trust taking ownership in 1999, the site had been neglected and it faced a huge task clearing the ash and dogwood to return the ground to chalk downland. Traditionally, sheep would have grazed the area and as part of a long-term management plan, it is hoped to reintroduce them in the near future.

In 1999, the Trust conducted an invertebrate survey of the small site, identifying 86 species. After five years of habitat management, a study in 2004 revealed that numbers of invertebrate species have doubled to 185, including two rare species – a weevil that feeds solely on hawthorn and an unusual hoverfly.

Saltbox Hill is a superb place to while away many hours exploring the habitat in search of flora and fauna. In the dew-soaked undergrowth, a snail descends a wet grass stem. Around the base of rotting trees are clusters of glistening inkcap. Unlike many autumnal fungi, the inkcap grows from late spring until early winter. Later in the day, the sun-drenched grassy hillside comes alive with butterflies; look out for the dark-green fritillary butterfly or the rare chalkhill blue.

Pyramidal orchid

Gone wild in urban South London, this enchanting cemetery provides the perfect environment to raise a family. Within minutes of closing time, a trio of month-old fox cubs emerge to play and sun themselves among the gravestones and fallen memorials. Once the gates are locked in late afternoon, the cemetery becomes an undisturbed sanctuary until morning.

Following a 53-day gestation period, the vixen gave birth to these cubs during mid-April in an underground den. Now in May, they are actively enjoying the world above ground. Even though it is spring, the late afternoon sun is still hot, so the cubs regularly seek shade in the ample vegetation. The thick undergrowth and headstones provide them with limitless escape routes should danger threaten. Safe and relaxed in their world, the cubs play, occasionally one wanders off, briefly following the dog or vixen as they set off to hunt. Perfect for exploration and games, the uncut grasslands are a favourite hang-out for hiding among the oxeye daisies, racing through meadow buttercups, climbing on the cemetery furniture or just play-fighting with each other. Occasionally a local dog barks, or a police car siren speeds along the nearby main road, causing the cubs to be momentarily alert, but they soon resume their carefree behaviour – their games pass the time between parental visits. An approaching adult is easily located, so often accompanied by a small party of mobbing crows.

Nearby a green woodpecker sits in an old tree – a haven for more than just foxes, many birds thrive in the mixed habitat. The woodpeckers prefer to feed in the areas of open grassland and enjoy the many anthills found around the gravestones. Kestrels and less common species such as spotted flycatchers have nested in recent years. Notable plants in spring include the delicate pink cuckoo-flower, swathes of cow parsley and four species of stonecrop, a small, succulent-leaved plant that grows well on the gravestones.

Through the evening and night, the tranquillity and wilderness feeling is broken only by the noise of a distant human world. But in the morning, it is time for the fox cubs to disappear from view. Soon a constant procession of commuters and excited school children will march through the grounds, oblivious to the activities of the cemetery's out-of-hours residents.

Probably the best-known wild resident of London, the fox has adapted superbly to life in the city, utilising the combination of man-made and natural habitats. Within the area circled by the M25 motorway, there are somewhere in the region of 9,000 to 12,000 foxes, though some fox experts believe that there has been a decline in numbers over the past few years. Inhabiting a wide variety of environments, from nature reserves to refuse sites, cricket grounds to back gardens, their lives intertwine with our own more than we may be aware.

If there was a favoured fox habitat within the city, I suspect it might be one of the unmanicured Victorian cemeteries, such as at Highgate, Brompton, Norwood or Nunhead. Normally locked at night, the cemeteries provide security, an oasis of biodiversity and easy access to the surrounding urban world.

Of huge importance to city wildlife, cemeteries are London's unofficial nature reserves. In stark contrast to the rapid development of city 'wastelands', graveyards that have been allowed to grow wild offer a positive change for wildlife that is rare. With enlightened local authorities, the 'grow wild' strategy often has official sanction. Indeed, Nunhead cemetery has become an 'official' nature reserve, jointly cared for by the community and local authority. Managed correctly, there is no reason why cemeteries cannot balance the needs of human visitors and wildlife. Ultimately, both have similar requirements – a place of safety and peace to be left undisturbed. Uncut wild flower meadows alongside cropped lawns, surrounded by native shrubs and trees offer a fitting tranquillity for visitors and a fantastic habitat for nature.

"...cemeteries provide an oasis of biodiversity"

Framed by headstones, the evening wanderings of a family of cemetery foxes are observed as intermittent sequences. On an evening in late spring, each short view allows a little insight into their daily lives.

UNDERGROUND, OVERGROUND – WANDERING FREE

Wimbledon's litter-collecting residents aside, London's railway routes offer a veritable haven for wildlife. The hundreds of miles of green corridors that criss-cross the capital provide valuable wild habitats and allow animals to travel relatively undisturbed. Apart from routine train passage, the only real intrusion many of these linesides receive is annual maintenance.

The rail tracks run through reserves of grassland, scrub and woodland or through wild parkland. Between Ravensbourne and Beckenham Hill stations are the wonderful old woodlands of Beckenham Place Park. Beautiful in any season, but particularly spring, the bisected Ash Plantation is a favoured breeding location for all three woodpeckers, parakeets and many species of warbler.

Travelling along one of the commuter routes to and from Central London is one of the easiest ways to view natural fox behaviour that is undisturbed by humans. Together with cemeteries and gardens, train lines are favoured fox habitat and are where they are most frequently observed. On some routes, a sighting is a probability rather than a possibility and at certain times of the year, more than one animal may be seen in close proximity. A few years ago, a fox family with cubs could regularly be seen relaxing among vegetation just outside Orpington station. Through the spring, they spent many of the days sunning themselves on a large decaying mattress, kindly donated by a fly-tipper.

Some of London's railway linesides are managed as official nature reserves, such as at New Cross Gate Cutting and Sydenham Hill Station. At New Cross Gate, a surprisingly wide strip of countryside conceals urban London from commuters for almost four kilometres, as they travel north from Forest Hill. On this stretch of track, tawny owls and six species of warbler regularly nest, including the undergrowth-loving whitethroat. To complement the rewarding train journey, three sections of the railway cutting are foot-accessible nature reserves, hosting among others frogs, newts, slowworms and lizards.

The banks of Sydenham Hill station are a reserve managed by the London Wildlife Trust. No direct access is possible, but the verdant habitat can be enjoyed from the comfort of a platform seat or from the entrance walkway and bridge. Sycamore, oak, hawthorn and a dense understorey of privet and elder sustain varied bird and insect life.

Not to be outdone, London's underground network also supports masses of wildlife – much more than just mosquitoes and tube mice. A recent survey by London Underground revealed more than 500 plant and invertebrate species growing alongside the tracks. How about encountering spider-hunting wasps, assassin bugs or fungus beetles on your commute to work? Or more appropriately, plants such as travellers joy, rat's tail fescue, speedwell, patience dock or just creeping-Jenny. Many birds, reptiles and amphibians were also discovered.

Along the trackside near stations there are badger setts and fox earths and at disused station tunnels in north London, roost sites of daubentons and pipistrelle bats. In total, twelve mammal species have been recorded, including stoat, common shrew and, most fittingly, the mole.

Railway lines are dangerous places and, apart from designated reserves, are not for exploration on foot. Let the train take the strain – enjoy the lineside safari from the comfort of a carriage and without disturbing the wildlife.

"Sweeping past city rooftops, mobs feast on swarms of airborne insects..."

Skies of screeching swifts signal the passing seasons, as spring speeds its sudden course to summer. Breeding visitors to the capital from sub-Saharan Africa, the first swifts normally arrive in the capital in late April, building in numbers throughout May. On dark sickle-shaped wings, the distinctive birds fill the air with their aerial acrobatics and magical sounds.

Entering the realms where other wildlife does not reach, swifts are amazing birds – almost mystical. They come and go with little warning and due to their aerial life, very few people have ever seen one close-up.

The swift's annual arrival in our skies is a cause for celebration. There are few natural spectacles that are as infectious, hypnotic and relaxing as the apparent delight of swifts, careering and reeling through the sky. Take a few minutes out on a warm afternoon or evening to watch them fly and you will be hypnotised.

Periods of calm gliding are interspersed with rapid beating of their thin wings. Sweeping past city rooftops, mobs feast on swarms of airborne insects, only landing momentarily to nest. Supreme masters of flight, swifts eat, drink, mate, bathe, gather nest material and even spend the night on the wing; climbing higher and higher into the dusk sky. With a few exceptions, they can even out-fly the most agile bird of prey.

Aptly named for more than one reason, swift perfectly describes the birds' super-fast flight and hectic life. A bird whose heart is firmly rooted in Africa, the swift's visit to London is short. Theirs is a race to incubate the eggs and raise one, two or three young before leaving the city, ahead of the summer.

In the City

Drenching themselves on the hottest day of the year, two city pigeons bathe in the cooling mist of a Trafalgar Square fountain. It is early morning and the day is already warm; by lunchtime, the temperature will reach the high 30s. In the summer heat, most bird activity is confined to the cooler hours around dusk and dawn. During the daytime, metropolitan monuments become the realm of flocking tourists, basking in the glorious weather. Discovering the sights and traditions of the capital, these summer visitors travel along established migration routes, through Trafalgar Square and its pigeon past, briefly setting down near Downing Street and on towards the city parks. However, it is not just tourists enjoying the splendour of London; some of the most visited landmarks conceal surprising modern-day residents. In the historic heart of the capital, there are rare birds breeding, secret wilderness areas and treasures along the Thames. Hop on board the alternative tour to the wild sights of the city.

Passionate moments on a hogweed flower for a pair of mating black-tipped soldier beetles. Early on a warm June morning, the umbrella-like flower heads play host to many pairs of the beetles. Just over one centimetre long, these summer-abundant insects are normally only observed when mating.

When hot days follow warm mornings and there is a profusion of insects, it is clear that summer has arrived. Bats are active over the city parks, busily gathering the insect harvest; about ten species of bats are known to occur in London.

The longer days offer the real chance of a late-evening encounter with a normally nocturnal badger. Attracted to gardens by regular feeding, the area around Croydon and Bromley is a stronghold for this well known, but little observed mammal. In mid-summer, the night is so short that badgers must emerge from their setts to start foraging well before dusk.

In summer, the city population swells with visitors. Feathered migrants such as

Poplar hawkmoth

"In Regent's Park a bee collects pollen from a perennial sow thistle..."

house martins and swallows explore the skies, while tourists explore the sights – different agendas, but often the same destination. Visited by millions for their architecture and history, London's landmarks can be some of the best places to discover wonderful plants and animals.

Even in the heart of the city, amid the imposing human world, there is a surprising diversity of wildlife. Cutting through the concrete are swathes of green – the Royal Parks, public and private gardens, land yet to be developed and cuttings along transport routes. This combination of open spaces and built-up areas gives London such a rich natural history. In Regent's Park a bee collects pollen from a perennial sow thistle and a hawkmoth rests on the trunk of a poplar tree.

Some species struggle to survive in this urban wilderness, but many, like the pigeon or peregrine, have turned it to their advantage. Once threatened with extinction, in 1962, there were only 68 pairs of peregrines across the country. Having since staged a remarkable recovery, peregrines are now the landmark bird of London.

Filling many guidebook pages, the landmarks along the redeveloped and tidied South Bank are popular destinations for tourists and wildlife! On the riverside path, midway between Waterloo and London Bridge stations, stands the old Bankside Power Station, now home to the Tate Modern. Waiting by the gallery entrance, an unwary visitor receives a splattering of bird droppings from high above. Wiping the mess from her shoulder, the woman cautiously glances upwards and curses the ubiquitous pigeons. However, her avian assailant is as complete a contrast to the abundant feral pigeon as London offers. The old power station's central chimney, shooting a mighty 100 metres into the city skyline is now a regular roost site for a pair of peregrine falcons, one of the capital's newest and rarest birds.

A bird of prey traditionally associated with cliffs and mountains, the peregrine is moving into cities, swapping natural crags and crevices for ledges and cavities on tall buildings. Perched just a few metres from the top of the Tate Modern chimney, this large falcon has a superb vantage point to survey its central London territory – taking in Shakespeare's Globe Theatre below and across the Thames on the Millennium Footbridge, to St Paul's Cathedral and further north to Regent's Park and the BT Telecom Tower.

With a preference for roost sites of notoriety, peregrines also use the Houses of Parliament and the Millennium Dome. The dome's towering metal supports provide perfect perches to watch for passing pigeons, gulls or other birds on which they prey. Masters of flight, peregrines hunt and catch their prey in the air. Partly closing their wings for speed they make spectacular aerial dives, known as stoops. A plummeting peregrine can reach an incredible 250 km per hour as it stoops down upon its victim. If successful, the bird will take its dead

prey to a high vantage point to pluck the feathers and begin feeding.

During the Second World War, carrier pigeons were used to transport messages; to protect these feathered messengers the government encouraged the killing of peregrines. In the 1950s and 60s the UK peregrine population plummeted, caused in part by the hunting, but also because they were affected by the overuse of pesticides, notably DDT. Only when a voluntary ban on such chemicals was introduced in 1962, did numbers begin to increase in their traditional rural habitats in Scotland, Wales and the West Country. In recent years, the species has begun to spread further east, entering terrain where it never bred before.

The peregrine falcon is one of London's wild success stories; four or five pairs now breed in the capital. With plenty of available prey and no shortage of nesting and roosting sites among the tower blocks and high-rise offices, there seems to be no reason why numbers should not increase. Hopefully the peregrine will become a familiar sight above the city and who knows, may even get its own entry in the tourist guidebooks.

"...a preference for roost sites of notoriety, peregrines also use the Millennium Dome and the Houses of Parliament"

"the world's fastest bird soared high above..."

In 2004, while queuing for Madame Tussaud's or The London Planetarium, early summer visitors were treated to breathtaking aerial displays, as the world's fastest bird soared high above the buildings. That year a pair of peregrine falcons nested nearby, at the top of the 22-storey University of Westminster building.

Nesting peregrines are continually threatened by the illegal activities of egg thieves – this pair's complete clutch was stolen the previous year to become part of a secret and very sad egg collection. So in 2004, the falcons received special protection from the Metropolitan Police, Regent's Park Rangers and local peregrine enthusiast Dave Johnson. From the time the eggs were laid in late March, Dave maintained a constant watch over the birds, clocking up hundreds of hours of observation to make sure that this year, no one spoilt the peregrines' attempt to breed. Three eggs hatched at the beginning of May, but a few weeks later only two chicks could be seen. From the remote watch point, it was difficult to ascertain what had happened to the third chick. The two surviving young were a male and a female. Female peregrines are significantly larger than males and can be easily identified at a distance.

Over the next six weeks, the youngsters developed from cute, but ungainly, fluffy chicks to impressive birds resembling their parents. By early June, the juvenile peregrines were vigorously exercising their wings – nesting so high, it is vital to get flying right first time. Flapping along the narrow ledge, they appeared less than secure and on reaching the corner would often look over the edge. These were nervous moments for the monitors and probably for the chicks themselves. Each time the adult female flew near, she would call, encouraging her young to fly. The excited and hungry young returned her cries. The cries of nesting peregrines are fantastic sounds when you know what they are. However, for local people unaware of the birds' rarity and existence, they were a noisy 5 am wake-up call, which continued every morning for a month.

On 14th June, the male flew for the first time, but mostly downwards. Unable to regain height, he ended up near to the treacherous main road. The only option was for Dave Johnson and Regent's Park Ranger, Tony Duckett, to catch him and return him to near the nest ledge.

In a more traditional nesting location, the juvenile may be able to make short upward flights, moving from one cliff ledge to another. The downside to this urban location is that there are no surrounding mid-height buildings to enable a gradual ascent.

Safely back on the ledge, following the brief human intervention, both the young fed on prey items brought in by the parents. Peregrines often cache food for consumption later. These adults had favoured 'larders' nearby – the male stored food at the top of the BT Tower. Subsequent examination of the nesting ledge revealed some quite unusual prey remains, from pigeons and starlings to parakeets, a woodpecker and even a domestic cockatiel.

Perching perilously close to the traffic streaming along Marylebone Road, this young female peregrine has just landed after her maiden flight. At 5.40 am, after much wing-flapping, she launched off the top of the tall tower block, as the first pictures of her flight show. Unfortunately, like her sibling, it was a mostly earthwards voyage, almost to ground level. Only five metres above the road, she must somehow return to the safety of the nest ledge, 22 stories above. With eyes tethered to the pavement on their daily journey to work, most commuters pass her by unnoticed. But her anonymity does not last long, as this historic event makes the afternoon paper and by the evening rush hour she is talk of the town.

During the next two days, both the juveniles made successful circling flights – taking off and flying to a neighbouring tower block before returning to their nest ledge. At six and a half weeks both young peregrines were flying competently, joining their parents for aerial acrobatics. Visitors to Regent's Park were treated to some amazing sights of central London's first peregrine family wheeling in the sky above.

Having mastered flight, the young left the nesting ledge completely and began exploring the city. The Tate Modern became one of their favoured roost sites and varying members of the family could regularly be seen resting on the tiny brick ledges, 100 metres up. On one occasion just after sunrise, both adult birds were present at the top of the chimney. Through the morning, the pair moved sides to avoid the advancing hot sun, until the adult male suddenly took off. At the same time, another peregrine could be heard flying in – the young male with a pigeon in its talons. As the two peregrines passed in mid-air, they called to each other. By chance, a flock of very panicked common terns also flew by at the same time. Surprised, the terns lost formation and for a few moments the air was full of confusion and the evocative piercing sounds of both species calling.

In 2005, the pair nested on the ledge again. The three eggs hatched in early June and, via a webcam, live pictures of the tiny chicks were broadcast on national television, enthralling millions of viewers. In mid-July the young flew for the first time.

Taking flight from Trafalgar Square, is it the end of an era for birds at this famous London sight? Once a tourist attraction itself, feeding the pigeons has recently been banned for health and cleanliness reasons. And to further encourage the pigeons to disperse, a falconer has been engaged to fly a hawk above the square – history has gone full circle.

In the 13th century, long before Trafalgar Square was even dreamt of, Edward I used the area as a place to keep and train his falcons. Later, the Royal Mews, which takes its name from the mewing call of birds kept for falconry, was the site of Henry VIII's stables. Then in the years just prior to construction of the square it was London's largest aviary. Exotic birds of all kinds – peacocks, lorikeets, parrots and even birds of paradise were kept in a vast menagerie enjoyed by its patron, George IV. Trafalgar Square was built in the 1830s and 40s and since then has been home to a massive flock of city pigeons, which are descendents of the coastal rock dove. For decades, tourists and Londoners have visited the square to enjoy the pigeon feeding frenzy – groups of birds taking bread or grain directly from their hands.

"...to feed, or not to feed? That is the question."

However, feeding the pigeons is a hot topic and has generated a long-running debate as to whether they should be fed – there are passionate views held by those for and against. Many people see the pigeons as an avian spectacle as much a part of the city as street theatre in Covent Garden; others believe the birds are simply a messy pest. To feed, or not to feed? That is the question.

In 2001, the Greater London Authority (GLA) ended the pigeon debate by making feeding illegal – this sparked a major row. Mayor of London, Ken Livingstone, believes the ban will counter years of over-feeding of pigeons and the reduction of pigeon numbers will bring a cleaner, healthier environment to Trafalgar Square. Recently pedestrianised to give greater public accessibility, the GLA hopes these measures will encourage tourists to enjoy this historical site and the outdoor café atmosphere (for humans)!

The abrupt end to feeding raised concerns that the pigeons would starve en masse. So, following lengthy discussions between the pro- and anti-feeding camps, a form of controlled feeding has been re-introduced. Every morning at 7 am, designated animal welfare volunteers are allowed to feed the birds for fifteen minutes. For a short while the square comes alive with the bird chaos and resembles its former self, but by eight, all bar a handful are gone. After the pigeons finish feeding, the square is hosed down to clean away the mess, leaving no evidence of the earlier avian spectacle.

With strong opinions on both sides, it is clear the feeding debate will continue for sometime.

Floral anarchy and natural disorder thrive amid the security of Downing Street. The violet hues of meadow cranesbill and tufted vetch bring vibrancy to the Cabinet Office wildflower meadow. Once the site of Henry VIII's tennis courts, this historical area has been seeded with wildflowers and is allowed to grow untamed through summer and then cut back in autumn. In one corner of this wilderness patch is a pond, its waters, neighbouring log piles and dense vegetation offer home to frogs. In mid-summer there is a wealth of insect life enjoying the flora – bees and hoverflies enjoy other flower species such as St John's wort, yarrow, nettle-leaved bellflower and cow parsley. Above the meadow on ivy clad walls are specially placed bat and bird boxes, which have recently been used by a pair of blue tits.

Whether famous or anonymous, small or large, city gardens provide pockets of green space among the jungle of concrete. Roof gardens and window boxes can, with the right planting, attract bees and other insects, bringing life to the urban world.

"the site of Henry VIII's tennis courts..."

Perhaps the most surprising residents of Downing Street are the pair of kestrels which have nested within a few feet of office windows at No.10. The previous year, the pair raised two chicks, but this year, there are four hungry mouths to feed. In the top of an old drainpipe, the nest site is very secure and the pair has little to fear from disturbance. Although, when the adults are away from the nest, many of the street's smaller birds pay significant if nervous attention to the kestrel chicks, regularly flying near the nest, almost mobbing them.

Behind the imposing black door of 10 Downing Street, the corridors of power lead out into the Prime Minister's garden. Here is a secret – a surprise wilderness area complete with wildlife pond, bird table and more.

Donated by the London Wildlife Trust and installed in 2000, the No.10 wildlife pond now holds good populations of aquatic life, from pond skaters, lesser water boatmen and freshwater shrimps to dragonflies and damselflies. In spring, the pond heaves with frogs and spawn, but how the adults reach the high-security habitat no one knows. The Prime Minister's young son, Leo, has a special interest in the garden's wildlife, particularly its mini beasts – keenly watching the tadpoles as they develop. Around the garden in the plant-covered walls many birds nest, such as wrens, robins, wood-pigeons, blue and great tits.

At such a famous address even the humblest wild visitor can become a celebrity. During the general election campaign of 2005, a pair of mallards nested on the lawn beside the pond. The female incubated 14 eggs for four weeks, completely oblivious to the status of her nesting location or the hype in the build-up to polling day. Thirteen of the eggs hatched and shortly afterwards mother and her train of little ducklings left the gardens. Treated like VIPs, the traffic was stopped and her family escorted across the road. Although Downing Street's secret wilderness areas seem to be surrounded by an impenetrable mass of imposing buildings, it is only a short waddle to Horse Guards Parade and across the road into St James's Park.

The Royal Park may be the realm of a captive collection of ducks, geese and pelicans, but it too has some wild surprises. Only a few hundred metres from the Prime Minister's official residence an unlikely reed warbler sings from the cover of waterside vegetation on the edge of St James's Park Lake. The warbler is just one example of the real wealth of the city's green spaces and particularly the Royal Parks that stretch for almost four kilometres west of the warbler.

*"Treated like VIPs,
the traffic was stopped and
her family escorted
across the road..."*

Oases of calm amid a thriving city, the four Royal Parks sweep through the centre of London forming an almost unbroken green corridor from St James's Park to Kensington Gardens. Originally enclosed as deer parks for hunting by Henry VIII in the 1500s, it was a hundred years later in the mid-17th century that the parks became open to the public. Today, this large stretch of undeveloped habitat supports an amazing profusion of wildlife, not normally expected in the heart of the city. Though chiefly recreational areas for public enjoyment, the Royal Parks are managed with wildlife in mind. Bat and bird boxes have been put up, areas of the parks are specifically designated for wildlife and each year many stretches of grassland are left to grow as wild flower meadows, to encourage insects and the many larger animals that feed on them.

Westwards from St James's Park is The Green Park. Without a lake, its habitat is very different from the other three parks. Dominated by mature trees and rough grassland, it is a superb place to look for the exotic-looking jay. One of the capital's normally elusive birds, their fruit and nut gathering antics can be observed at surprisingly close range. Birds and animals in The Green Park benefit greatly from the wildlife-friendly gardens of their neighbour – Buckingham Palace. Her Majesty The Queen's gardens at Buckingham Palace have some

ıabitat and are continually being ildlife. Around the Palace's large ıg grass have been left ffer cover for insects and other ́all bay trees provide a nesting wls and standing deadwood ɔdpeckers. Purposely placed log ɜr for amphibians and other ɜs stag beetles.

west through the Royal parks ɘ relentless traffic flow at Hyde ıjects a brief intrusion in the tranquillity, before entering into 250 hectares of calm in Hyde Park and Kensington Gardens. With their grounds and waters continuous, Hyde Park's The Serpentine becomes The Long Water of Kensington Gardens under the West Carriage Drive Bridge. Herons are easily located residents of these waters and regularly can be found perching on the various lake-edge posts. Great crested grebes often nest within sight of the bridge.

The Long Water is a regular haunt of cormorants and black-headed gulls. Mallard, pochard and tufted duck breed on The Serpentine and its concrete edges can be good places to watch moorhen and pied wagtails. House martins and common terns circle above the lake, looking for fish or summer insects. The park's many small copses, ornamental gardens and grass areas support a host of traditional garden birds. Further west, mute swans swim gracefully across the Round Pond in Kensington Gardens; in abundance these elegant birds dominate the vista.

"Terns arrive in Hyde Park from their wintering grounds on the coast of West Africa..."

Skimming along the Serpentine, the sight of common terns fishing the waters of Hyde Park is one of summer's many pleasures. Silvery-white with a black cap and red bill, the elegant terns have long streamer-like tails; because of which they were once known as 'sea-swallows'. Most years, a couple of pairs arrive in Hyde Park from their wintering grounds on the coast of West Africa. Throughout the day they cruise up and down the edge of the Serpentine and adjoining Long Water of Kensington Gardens. They are easily located by their shrill 'kee-yaa' call. Systematically scouring the lake edge, the terns' flight is graceful and considered as they search for small fish. Once prey is sighted, they may hover briefly, before plunging into the water. Hopefully, they emerge from the ripples with a fish, but not always.

An eclectic creation, constructed from fluorescent Frisbees, feathers, shiny litter and sticks, this distinctive nest has been built by a pair of coots. Woven together on top of a bright orange life-saving ring, the unique nest brings a splash of colour to the Round Pond in Kensington Gardens. For a decade this pair has nested on the large pond and each year, their nest design evolves dependent upon the material available. Seen here in August, the coots are incubating their second clutch of eggs.

One of the most abundant waterbirds in London, coots are very protective of their territory and will confidently chase off other coots and species larger than themselves.

A nest with palatial views: this great crested grebe has Kensington Palace and the coots as neighbours. Its nest is a more subtle affair, merely water plants draped over a collection of sticks. Once the young hatch, they are often carried on the backs of the parents. Unlike the adults, the chicks are striped white and black with a deep red spot between the bill and eye. The young are fed insects and small fish, such as the three-spined sticklebacks found in the Round Pond. By six to eight weeks, the young have lost much of their stripy appearance and within another few weeks will be independent of their parents. The waters in Kensington Gardens and Hyde Park are fantastic places to watch grebes nest and raise young at close range. The first great crested grebes to breed in central London nested here in 1972.

The central London green corridor ends at Kensington Palace, although only a few blocks west are the leafy glades of Holland Park or further south across the Thames is Battersea Park. With a heronry, river wildlife and a nature reserve managed by the London Wildlife Trust, Battersea Park has quite a diversity of flora and fauna. Along the route south from Kensington Gardens are two very different wildlife giants – a museum and a power station.

left: *Juvenile great creasted grebe*

Stuffed with wildlife, the Natural History Museum seems a logical place to find nature in the city. And though most of its specimens are long dead, concealed in a corner of the museum's grounds is a vibrant and flourishing wildlife garden. Built as the Museum's first living exhibition, the garden represents eight different habitats, ranging from reed bed to chalk downland. The centrepiece is a large pond decorated with a profusion of lily leaves – perfect habitat for moorhens, dragonflies, damselflies and frogs. Foxes, bats and squirrels are regular visitors, the latter enjoying nuts of the hazel coppice in late summer. More than fifty species of birds have been observed from the garden, many nesting within its boundaries. Perhaps most exquisite is the nest of a pair of long-tailed tits, who construct an enclosed oval ball from moss, hair and cobwebs, then line it with feathers.

Hoverflies feed on the summer nectar bounty provided by the flowers – this species, *Volucella inanis*, rests on a wild carrot flower head. More than 300 moth and butterfly species have been recorded in the garden, including a flourishing population of six-spot burnet moths. Speckled wood butterflies are exciting to watch as they battle for the sunniest spots in the woodland. In sunny glades, males rise into the air and perform spiralling duels, sparkling in the rays of sunlight. Eventually, the winner settles on a favoured leaf – the prime site to attract female attention.

Hoverfly

Speckled wood butterfly

"...less than a hundred pairs of black redstarts breed in the UK"

Now merely a shell, two of London's rarest bird species breed in the emptiness of Battersea Power Station. Lying disused and awaiting redevelopment, the building is a striking structure dominated by four chimneys – a distinctive landmark on the south bank of the Thames.

Among the rubble and brick foundations are breeding black redstarts. Less than a hundred pairs of the redstarts breed in the UK, making the species one of Britain's rarest birds. Greater London is its stronghold. Black redstarts need sparsely vegetated rocky areas with suitable singing posts and so derelict urban land offers prime habitat. Unusually, the Battersea birds are resident year round, whereas redstarts elsewhere in the city move to favoured wintering grounds – often at sewerage farms such as Beddington or Barking. Probably because of its size and riverside location, the power station offers a healthy insect population throughout the seasons.

Black redstarts are relatively new to London – they first bred in the capital in 1933. Once known as 'bomb-site' birds, the redstart population increased significantly after the Second World War because of the sudden profusion of suitable habitat. The Blitz created perfect redstart terrain and for a while their numbers grew.

However, today, the rapid redevelopment of the city is causing the black redstart to lose ground. How Battersea power station is redeveloped will determine whether the resident redstarts survive. Thankfully, there is a good awareness of the birds' predicament and many new building projects incorporate their needs into the development. Living roofs can return some of the space lost by the construction and the provision of suitable singing perches also helps. Further east along the Thames, the area around Deptford is a good example of how the redstarts have helped shape redevelopment.

Battersea is also known for its breeding peregrines. 2003 saw a landmark event for London when a pair of peregrines successfully bred and raised young at a nest site on the power station. Of the three chicks, two died during early attempts at flying. However, one successfully fledged, leaving the nest for the last time in June.

Due to imminent construction work on the power station, a lofty tower was built nearby to encourage the birds to breed away from the structure in future years. So far, the peregrines have ignored this new home and have continued to nest on the building itself. In 2005, the pair successfully raised four young.

Monuments to modern urbanity, the great towers of lifeless glass and concrete around Canary Wharf dominate the skyline. From the derelict docklands area, this high-rise mini city has risen and is now home to many of London's tallest buildings. Yet even here there is wildlife. Peregrines have been seen roosting in the area and there is hope that they may breed on these contemporary cliffs. There are no canaries, but it is easy to find little grebe, cormorant and lesser black-backed gull on the abundant water. On summer days, the sight of common terns fishing in the docks is spectacular – images of plunging terns reflect from the glass sides of surrounding buildings. Attracted by the good feeding, the terns nest on specially placed islands in some of the surrounding docks.

On both sides of the river, the vast docks were once the centre for the capital's trade, from whaling and timber to sugar and tobacco. With containerisation and mechanisation, the trade moved outside London to ports such as Tilbury. The area was in decline and by the late 1960s, most of the docks had closed and the area became derelict. However, over the past couple of decades a massive regeneration programme has taken place and Docklands has undergone a dramatic transformation. Many of the old warehouses have been converted into apartments and new nature reserves have emerged alongside tower blocks.

Lavender Pond Nature Park in Rotherhithe is one of the oldest urban nature reserves in the UK and its pond and adjoining pumphouse reveal a little of the history of this area. A part of the old Surrey Commercial Docks – the centre of Britain's timber industry – Lavender Pond was built in 1815 as a shallow expanse of water where timber was floated to prevent it cracking or drying out. In 1970, as the demand for traditional docks declined, Surrey Docks was closed and Lavender Pond filled in.

In 1981, Lavender Pond was brought back to life as a nature park, complete with open water, marshes, reed beds and the renovated pumphouse as an educational museum. Reed warblers now nest in the reeds; swans, moorhens and ducks also breed here. Raised boardwalks lead visitors through the dense reed beds offering glimpses of the furtive warblers or longer views of resting damselflies.

Today, the old Surrey Docks site is also home to Russia Dock Woodland and Stave Hill Ecology Park. Stave Hill is a montage of different wild habitats, from wet marsh and wildflower meadow to young woodland. Standing on top of the hill there are stunning easterly views over Docklands and westwards towards central London.

"even here there is wildlife..."

Lavender Pond and Stave Hill are managed by the Trust for Urban Ecology (TRUE), which also manages Greenwich Peninsula Ecology Park and Dulwich Upper Wood. Formed in 1976 to develop creative conservation in urban areas, the trust established Britain's first urban ecology park. (The temporary William Curtis Ecological Park was established on a derelict lorry park near Tower Bridge.) When the site was handed back to its owners in 1985, TRUE took on the Stave Hill site as a permanent replacement. Just twenty years old, Stave Hill boasts some impressive wildlife, not least of which is an established colony of the protected great crested newt. More than twenty butterfly species can be found around the site, which is Britain's first urban butterfly sanctuary. Careful exploration of the grasslands might reveal slowworms, frogs, toads, or perhaps hedgehogs, while enjoying the beautiful song of skylarks overhead and even a nightingale singing from among the scrub!

The Millennium Dome has settled into a sedate existence on the northern waterfront of the Greenwich Peninsular. In its formative years, spiralling costs and construction delays caused uproar and brought loud protests about its existence. However, with the year 2000 long gone, the Dome has become a mature landmark, only occasionally stirred by rock concerts or similar events. Surrounded by some inspiring and innovative wild habitats, the Dome has become a favoured site for falcons – kestrels and peregrines regularly sit on its towering structure.

Known as Greenwich Marsh, the peninsular was once agricultural field and marshland. Then in the late 19th century, the area was heavily developed with the construction of the gas works, a shipbuilding yard and the Blackwall Tunnel. By 1968, there was no green space left on the peninsular except for the British Gas playing field.

Then in 1997, English Partnership bought the 121 hectares of peninsular land and began a massive regeneration project. As well as the Dome, it also planned for nature, establishing pocket wild habitats and restoring the riverbank, terracing it in parts to allow plants such as club rush, sea aster and common reed to establish. This stretch of the Thames Path is fabulous to walk in summer and allows access to most of the waterfront. Starting in the west among the last of the peninsular's industrial area, the path travels through black redstart territory before reaching the Dome. Following the water's edge around the Millennium Dome, you pass Ordnance Wharf, a redundant jetty that has been allowed to become colonised by plants and wild flowers like a living roof. At the base of the Dome behind security fencing are the Meridian Gardens, an area of wild flower meadow and reed bed. On the river foreshore at low tide, herons, gulls and cormorants gather on the sandy beach. Further along, the water's edge is terraced and the reeds are home to many pairs of reed warblers – difficult to observe but wonderful to listen to. Swans, shelduck and other wildfowl frequent the low water mudflats; less obvious are waders like little ringed plovers, though they do come close to the pathway.

By far the most impressive feature of the area is the Greenwich Peninsular Ecology Park, a large freshwater habitat overlooked by the award-winning Millennium Village. Opened in 2002, within just a few years the park has taken on the appearance of an established habitat. Two large lakes are edged by great swathes of reed bed and areas of wet woodland – both willow and alder carr. Raised boardwalks trail through the park to two hides that allow disturbance-free viewing of the wetland. Common terns and moorhens nest in the middle of the lake on artificial shingle-covered platforms. The hides offer the best views of nesting terns in London and also some of the closest reed warbler sightings. Both reed buntings and warblers nest in the park, as do little grebes and the more common duck species. In summer, particularly before rain, the lakes attract house martins and swifts, skimming along the surface to catch insects. Flying so low, the swifts provide observers with outstanding views, often sweeping past within a few feet.

Florally, the park is superb in summer. The wild flower meadows bring vibrant colour and wetland plants such as yellow flag iris and water lily provide a striking focus. The most unique feature is the long stretch of shingle beach that runs parallel to the neighbouring Thames Path. Bare in winter, the beach becomes awash with flowers in spring and summer. Wild carrot, yarrow, bird's-foot trefoil and meadowsweet erupt from among the rocks and pebbles, attracting numerous insects and butterflies. The bird's-foot trefoil in particular is the caterpillar food plant of the stunning black and crimson six-spot burnet moth.

"...the beach becomes awash with flowers"

A migrant hawker darts through the marshland vegetation – bird hides allow very close views of reed bed activity. In its opening year, 14 species of dragonflies and damselflies were recorded at the Greenwich Peninsular Ecology Park; the reeds provide perfect habitat for them. Many other insect species inhabit this superb wetland habitat. On a morning in late summer, a field grasshopper rests on some long grass.

Dotted with urban jetsam, Deptford Creek seems an unlikely wild location – at low tide the steep concrete sides edge an unpromising muddy channel. However, the creek is a unique brownfield environment and deserves really close inspection. For grey wagtails and black redstarts, the recently refurbished creek offers food, water and plenty of secure nest sites within its walls and bridge arches.

Deptford Creek is the tidal section of the Ravensbourne River that flows, covered in parts, from Keston in south-east London through Beckenham Place Park and Lewisham until it reaches the Thames at Deptford. Once

LIVING ROOFS

Living roofs are a beautifully simple concept – encourage new wild sites to grow on the top of buildings, to replace ground-level habitat lost through development. Although not a like-for-like swap, with the emphasis on urban regeneration, living roofs can help to mitigate a development's impact on the environment.

The construction of a living roof involves covering the surface of the roof with a special waterproof membrane and then placing soil or a similar growing medium on top. The 'soil' is seeded, planted or allowed to colonise naturally with vegetation. Varying slightly from this 'green' roof is the 'brown' roof, where the surface is covered with crushed rubble from local sites and then left for invertebrates and plants to move in slowly.

Thankfully, through the co-operation of developers and conservationists, living roofs have already been installed on a significant number of sites, particularly in east London. Great for wildlife, building insulation and reducing air pollution, living roofs are planned for many new developments.

The Creekside Centre and Laban Dance Centre both have brown roofs to replicate the local brownfield habitat. This brownfield land is the preferred terrain for black redstarts, but as derelict urban sites are rapidly being redeveloped, these rare birds are losing ground. If the black redstart is to have a chance of surviving, brown roofs will play an important part.

Nearby at the Greenwich Peninsular Ecology Park, the visitor centre has a verdant green roof, bedecked with stunning wild flowers and occasionally a pair of Canada geese. On a living roof in Canary Wharf, researchers recently found a new species of spider for England.

a major dumping ground for rubbish of all varieties, the creek underwent a significant clean-up with the arrival of the Docklands Light Railway. Since 1996, local people have performed an amazing task clearing the refuse. Literally hundreds of shopping trolleys have been extracted from the mud, while tyres, street cones, industrial waste and other undesirables have all been removed too.

The transformation was completed in 2002 when the Creekside Centre opened. Topped with a living brown roof, the centre is a place where schools and other groups can learn about the local environment. Having donned a pair of waders and with the expert guidance of the Creekside staff, visitors really can immerse themselves in this intriguing habitat.

Explored at low water, the industrial landscape reveals some surprising residents. Search the shallows or pools along the creek bed and there are freshwater shrimps and crabs hiding under rocks and concrete boulders. The Chinese mitten crab introduced from the Far East a century ago has firmly established itself in the Thames over the past two decades. A burrowing species, there are worries that the mitten crab may cause serious bank erosion. Along the creek are the golden yellow flowers of hawkweed oxtongue, on which a nationally rare beetle, *Olibrus flavicornis*, breeds.

Wading downstream towards the Thames, ahead is the award-winning Laban Dance Centre, which was also built with a living roof – one of the first in the capital. The creek flows under one of the earliest railway viaducts in the world. Dating from 1834, the brick arches were built for the Greenwich to London Bridge railway. With every clumsy stride the visitor makes, small fish dash across the creek bottom, leaving a cloudy trail, slowly merging to render the migrant invisible. These are flounder fry, visitors to the creek through the summer. Their migrational movements are intricate and the creek so important to them.

"…as the flounder grow, the left eye will migrate towards the right side"

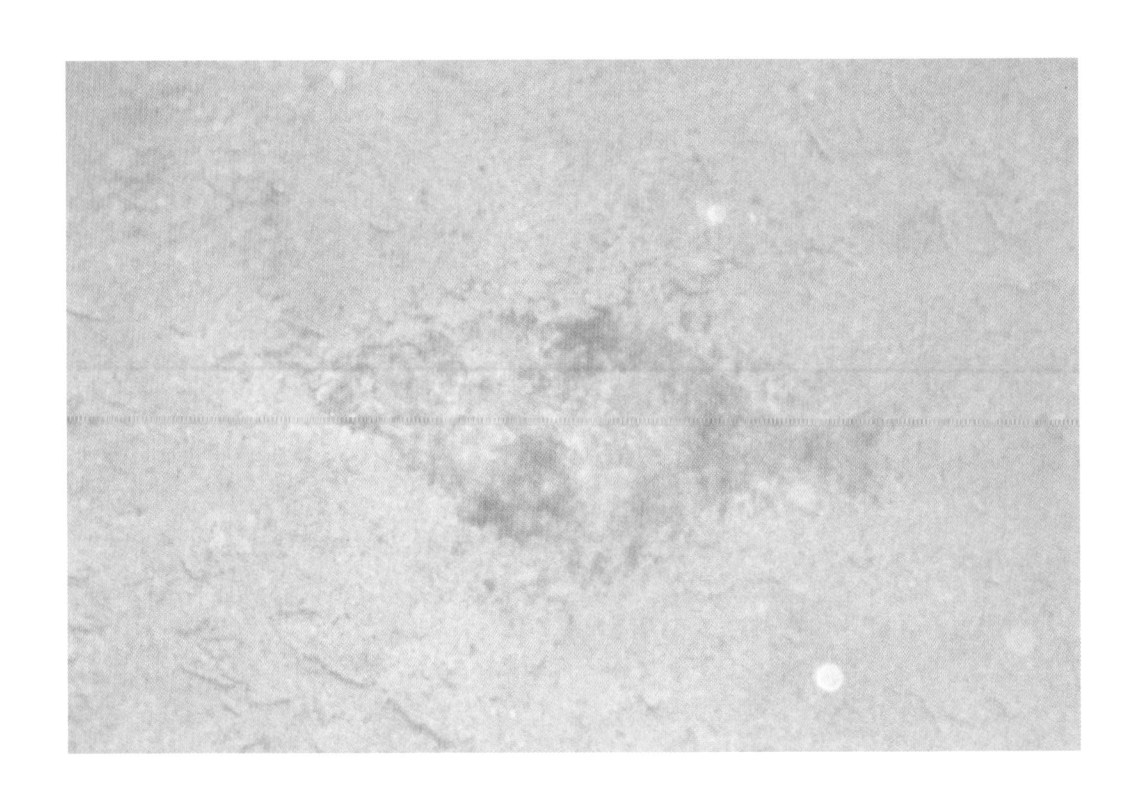

The Thames Estuary is a key nursery ground for many species of fish and is a spawning ground for flounder. Flounder are a type of flatfish, which because of their riverbed lives have both eyes on the upper side of their body. However, the fry (newly spawned fish) appear like other fish with eyes on both sides of their head. Amazingly, as the fish grow, the left eye will migrate towards the right side, hence their being named right-sided flounder.

Within days, the thumbnail-size fry must move towards their growing grounds, of which Deptford Creek is one. Not yet able to swim against strong currents, the fry use tidal movements to migrate. The young fish rise up in the flood tides and take refuge along the river margins in the ebb tide, slowly progressing upstream. In Roman times, the Thames was much wider than it is today and, subsequently, the river flow is now much stronger. As riverbank development continues, it is becoming increasingly harder for fry to fight the strong currents to complete the migration.

Assuming the fry have survived the journey, by spring they will have reached the relative safety of Deptford Creek. Here the flounder fry will stay to grow through the summer months. In the autumn, they must return to the sea, drawn by the warmer and more saline waters – avoiding predatory cormorants as they travel.

Seeing skies abandoned by Africa-bound swifts, leaves a sinking feeling that summer will soon be over. But it is only early August, the peak holiday season and the hottest days are still ahead. Though London's landmarks still teem with tourists, the first signs of autumn are showing. Thankfully, most summer migrants will stay in the capital until September, but swifts and cuckoos are already heading south.

With no parenting responsibilities, adult cuckoos can leave very early. Reed warblers and the other host species have the job of raising the single cuckoo chick. Having ejected the eggs of its host early on, the young cuckoo grows fast and will be ready to depart a few weeks after its parents. In late July, this juvenile in Rainham Marsh is already flying and will make the trip back to Africa independently of its parents – an incredible example of a bird's inbuilt navigational ability.

With somewhere to be, the cuckoo and swift are on a strict timescale, but there is one city sight they will not have seen. Hot on the heels of the Notting Hill Carnival is the riotous colour festival known as autumn.

"...cuckoos are already heading south"

Colours of London

Displaying his battle colours, a mud-darkened red deer stag bellows across a field of browning ferns in Richmond Park. To enhance his grandeur, thrashed bronze bracken adorns his already impressive antlers. Autumn heralds the onset of the annual rut, when mature males must compete for territory and females. Initially, the contest is by sight, sound and smell, but if stags are equally matched, physical sparring and locking of antlers will take place. Rivalling the stag's rutting spectacular are the city's trees and woodlands. As London's leaves change from gentle greens to glorious golds, an arboreal masterpiece evolves. Riotous red and orange berries decorate the hedgerows, presenting a seasonal banquet for birds and mammals. And though shorter days and the first frosts will slowly drain colour from the vegetation, the cold weather gatherings of birds will bring a renewed feathered vibrancy to the capital.

Autumn is a time of great change and contrasts for wildlife in the capital. Trees, hedgerows and many of their inhabitants are at their most energetic and beautiful, though other species are already winding down or preparing to leave the city for warmer climes.

Having raised young, house martins can be seen gathered around their nesting sites or in rows along telegraph wires, boosting their energy supplies with frequent fly-catching forays. They will soon depart on an epic journey to their African wintering grounds. Good places to watch these pre-migrational congregations of martins are along the Thames and at the Chase Nature Reserve or Fairlop Waters in East London.

September is the peak month for summer breeding visitors such as terns, swallows and warblers, to move south. Among the towering offices at Canary Wharf, a tiny chiffchaff searches for insects on a red oak. While most visiting warblers will soon leave London for the warmth of Southern Europe or Africa, many chiffchaffs choose to remain in the capital.

Always a few degrees warmer than the surrounding countryside, London's artificial climate encourages some traditionally migrational species to remain resident. Human habitation and exotic flora offer an unseasonal food supply and our waste provides a regular insect yield. Chiffchaffs, black redstarts and blackcaps are bird species that are now found year round in the capital – though the blackcap's story has an unexpected twist. Increasing numbers of wintering blackcaps were, logically, thought to be summer birds staying year round. However, the leg-rings of 'ringed' wintering blackcaps show they are migrants from Germany and central Europe. It is thought that, perhaps by accident, some European birds flew north instead of south and ended up in the UK. In the urban and suburban warmth, the blackcaps have thrived through the cold spell, even visiting bird tables and feeders. The huge bonus for these birds is that they are closer to their summer breeding sites than the southerly wintering blackcaps and so can return earlier and select the best locations to nest.

This is a recent phenomenon and it is not known whether it will continue long-term. A greater study of the over-wintering summer migrants is needed before the change in migrational movements is fully understood.

Chiffchaff

The brown rats of central London may not rely on the autumnal food abundance, but some seasonal seeds are a healthy alternative to their normal diet of processed lunchtime scraps. However, autumn is the time for most mammals to build up reserves. Small rodents will certainly be enjoying the temporary glut of nuts and fruits. Few animals are more identified with foraging for fallen nuts than the grey squirrel. Throughout autumn, they busily search for acorns, hazelnuts and more exotic bounty to be found in London's parks. Squirrels must work fast to collect the spiny sweet chestnuts, as they are prized by another of the city's mammals. Delicious roasted over an open fire, chestnuts are a winter luxury that perhaps briefly remind us of our own species' gathering and self-sufficient past.

On a bright October day, Highgate Wood is a popular place, alive with joggers, families, dog-walkers and busy grey squirrels. Each of the oaks in Highgate may produce up to 50,000 acorns; a bounty for squirrels and jays who both horde the nuts. Stored for leaner times over the winter, buried caches of acorns will hopefully be found and retrieved later on, but those that remain undiscovered may germinate and grow into seedlings in the spring.

In early autumn, foxes are busy defending or establishing their territories prior to the mating season. Summer fox families have begun to break up and the young adults will be looking for a patch of their own. They may take over the range of dead or weak parents.

For reptiles and amphibians, the cooler days of September and October indicate that it is time to enter hibernation. Toads seek out unused mammal burrows or gaps underneath outbuildings, in which to hibernate through the winter. Frogs simply bury themselves in the mud and decaying vegetation at the bottom of ponds or climb into compost heaps or under old tree stumps.

Hidden under an old sheet of corrugated iron, an abruptly exposed slowworm retreats into the thick meadow grass. This slowworm is absorbing the warmth from the metal on a sunny late September day. Not a worm at all, these legless lizards are seldom seen, being most active at night. Found in many habitats across the capital including gardens, the reptile's penchant for slugs and snails makes it every gardener's ally. From mid-October, slowworms hibernate, often communally, among tree root crevices or in piles of leaves.

Rather than a decline towards winter, autumn is the grand feast, the finale that the rest of the year has been building up to. Parks and gardens are ablaze with fruits and berries of all colours, catering to many different tastes. In Kensington Gardens and Hyde Park, London's wilder residents are beginning to feast. Across the road, resplendent revellers pour out from the 'Last Night of the Proms' at the Royal Albert Hall, high on pomp and tradition. By mid-September, the internationally renowned Prom concert season has reached its crescendo, whereas the capital's trees and shrubs are just starting to stage their seasonal performance.

At the autumnal feast, the horse chestnut is one of the first to impress. Homeward promenaders may pass trees whose fruits have just begun to ripen revealing a glimpse of the shiny brown conker inside. Avidly collected by schoolchildren, the fallen nuts are threaded on to strings to play conkers. The winner or 'conqueror', from where the game's name is derived, is the one who has literally smashed the opposition. At the same time as the conkers are opening, the leaves of the horse chestnut have begun to turn golden. The outside rim of the large leaf changes colour first, the golden band slowly creeping inwards.

Hillingdon's hedgerow harvest blushes with a profusion of scarlet berries and golden leaves. Along the Yeading meadow, the blackthorn is an impressive sight. Large impenetrable thickets of blackthorn don their blue grey berries, while nearby the deep shocking pink fruits of spindle trees contrast wildly. Spindle fruits are inedible to humans, but sloes, the fruit of the blackthorn, are traditionally used to make jam and flavour gin as well as being enjoyed by the local fauna.

Laden with bright or orange red berries, blackbirds and thrushes adore the firethorn or pyracantha. A popular garden shrub, the pyracantha can tolerate the high atmospheric pollution of cities. Also prized by many different species are the profusion of succulent sugar-rich berries on trees of the *Sorbus* genus, relatives of the rare wild service tree. Enjoyed by wintering fieldfares, redwing and other thrushes, the fruits of whitebeam or rowan, also called mountain ash, are a vital food source and bring vibrancy to the hedgerow.

The glossy dark red, almost black, bouquets of elderberries are eagerly sought by birds and also made into jam and wine. Elder is a nitrogen lover and the trees flourish in many human habitats – in cemeteries, abandoned houses and where organic waste matter has enriched the soil such as around rabbit warrens or badger setts.

Perhaps the most prized and distinctive berry of late summer and early autumn hedgerows is the blackberry, which grows readily in most places, even on less fertile ground. Many Londoners will have fond memories of 'brambling' along roadsides, probably devouring more blackberries than make it home for pie or jam making. Feeding on blackberries from the hedgerow is one of those activities that provides so much pleasure.
But blackberries are also an important food for birds, small mammals and squirrels and perhaps surprisingly, foxes. Badgers and deer are also partial to these accessible delicacies.

"...a profusion of scarlet berries and golden leaves"

As summer departs, each tree leaf is responsible for the largest evolving mural in London. From September to November, the intoxicating arboreal art evolves through an extravagant array of shades and hues until finally being torn apart by winter winds. From a distance, the view can be stunning, but close-up it is simply magnificent. Understanding the choice of pigments used enhances the visual delight.

The leaf is a tree's workshop. Throughout the growing season of spring and summer, the leaf absorbs carbon dioxide from the air and takes in water drawn up from the roots. Chlorophyll – the green pigment dominant in the leaf – traps sunlight and uses this light energy to convert the carbon dioxide and water into carbohydrates or simple sugars. This is photosynthesis and it provides the sugar-rich sap essential for growth of the tree.

The leaf also contains other chemical compounds to protect the chlorophyll from excessive sunlight. The most common are the deep orange-coloured carotenoids and the yellow flavonoids.

Shortening days signal the arrival of autumn and it is less sunlight and not cold weather that causes the leaves of deciduous trees to change colour and fall. As sunlight decreases, trees slow down their growth and begin to reabsorb the green chlorophyll for use the next spring. This process gradually uncovers the yellow and orange pigments and, in turn, as they are each reabsorbed so the leaf slowly changes. The most arresting autumnal displays follow long or hot summers, when greater amounts of the golden protecting pigments have been created.

After absorbing these pigments, some trees, like oaks, release by-products of the summer, including tannins into the leaves, which give a brown colouration.

Finally, the tree develops a brittle cork-like seal at the base of each leaf and with the next strong breeze the foliage falls to the ground to decompose, its job done.

"…it is less sunlight and not cold weather that causes the leaves of deciduous trees to change colour and fall"

Looking across ancient oaks towards the historic skyline of central London, views do not come much finer. From Richmond Park, many of the capital's landmarks are visible, including St Paul's Cathedral, the London Eye and the most recent addition – the Gherkin.

Almost five per cent of Greater London is woodland, about a third of which is classed as ancient, having been continuously wooded since 1600 AD. Some of the oaks in Richmond are thought to be about seven hundred years old. The most impressive woodlands in the capital are found around its outer edges.

In the north-west of London are the beautiful beeches of Bayhurst Wood and coppiced hornbeams in Mad Bess Wood. Slightly south stand the ancient oaks of Ten Acre Wood. To the east is Epping Forest, a former royal hunting area that extends northwards for 12 miles from Wanstead Flats to beyond the M25 motorway. Nearby is Hainault Forest, where there are plans to restore adjoining farmland to its original glory as a mixed wood.

The greatest proportion of the capital's woodland is to be found across the river in the south-east, particularly between Croydon and Bromley. The borough of Bromley contains almost a quarter of London's woods, ranging from the sycamores and sweet chestnuts in Scadbury Park to the substantial stands around Charles Darwin's house at Downe. Heading north towards the centre of London, there are some remarkable woodlands – islands surrounded by a sea of houses. Of particular note are the woods of Dulwich, Sydenham Hill and Oxleas. Dulwich Upper Wood and Sydenham Hill Wood are surviving tracts of the old Great North Wood. Now nature reserves, both are a strange mixture of ancient woods, the remnants of Victorian gardens and new woodland. The ancient woodlands at Oxleas were the focus of a high-profile and successful conservation campaign in the

1990s. Threatened by road development for the East London River Crossing, it is thanks to intense local and national campaigning that this beautiful wood still stands today.

Jump on a tube and there are many wonderful woodlands within easy reach of the heart of London. Situated between Kensington Gardens and Olympia Exhibition Centre, Holland Park proudly flaunts its autumn glory and is bustling with wildlife. The subtle seasonal shades of the formal gardens and untamed woodland complement each other. Among the vegetation, songbirds flit from branch to branch, while below, unusually dark rabbits forage on the ground. Along Holland Park Avenue, the fallen leaves of plane trees are imprinted on to the edge of the busy road. Some of the plane trees in London's squares and along its city streets are more than 200 years old. The London plane is one of the few trees that could cope with the pollution of the capital prior to the Clean Air Act coming into force. A hybrid between the American and Eastern plane, the London plane has shiny leaves that wash clean in the rain, while its bark peels away in large patches – shedding the city grime. Sadly, the plane is not favoured by birds or insects and so does not support a great biodiversity.

To the north, Kenwood, Highgate and neighbouring Queens Wood all offer fantastic, but unique wooded habitats. Places that within minutes of entering make you feel you are far away from the noise and stress of urban London.

Highgate Wood was part of the Bishop of London's hunting estate in the medieval period. From the 16th century until the 1880s it was leased to tenants who harvested wood by coppicing the hornbeams. Then fearing it might be sold off for development, Henry Reader Williams headed a campaign to protect the wood. In 1886, it was gifted to the Corporation of London, which declared it open for public use forever. For magnificent trees that are easily accessible, Highgate is the place. Now a Site of Metropolitan Importance, the 28-hectare wood is a wild oasis, only four miles from central London. Almost a thousand species of invertebrate have been found within the wood, and more than 250 species of moth. The wood owes its fantastic insect biodiversity to the profusion of ancient trees and the practice of leaving fallen branches and logs to decay naturally. The arboreal heritage is also responsible for the seven species of bats recorded here.

London's first National Nature Reserve, Ruislip Woods was officially notified in May 1997. The reserve covers almost 300 hectares and comprises five areas, four broad-leaved woodlands –Mad Bess, Bayhurst, Copse and Park Wood, and Poor's Field, an area of common. At 100 hectares, Park Wood is the largest unbroken woodland in the capital. It is dominated by oak and hornbeam – both the English and sessile oaks are found here. The deciduous woodlands attract two of Britain's most impressive butterflies – the white admiral and the rare purple emperor. Park Wood is now separated from the rest of the reserve by a reservoir, Ruislip Lido, which was built in 1811 as a feeder for the Grand Junction Canal.

Exploring the largely oak and hornbeam woods, the most obvious features are the coppiced hornbeams. Until the coppice industry collapsed in the 1930s, these woods would have been managed as coppice with standards. This is where the hornbeam is regularly cut to a stump, encouraging new growth, which is then harvested as poles or for fuel, with the surrounding oaks (standards) allowed to grow to maturity. The varied tree growth in a coppiced wood supports a wide diversity of wildlife and so over the past two decades, coppicing has been reintroduced to regenerate the woodland.

"...the old woman prowled the woods at night looking for poachers"

A striking feature of any deciduous woodland in late autumn is the many hued leaf litter, which draws attention to ground level. Kicked into the air by playing children, the fallen leaves conceal a world of incredibly diverse invertebrates. The disturbed litter offers easy foraging for birds such as robins.

Squirrels dash across the forest carpet busily gathering acorns and beech mast; unique to Bayhurst Wood, the beech does not occur in the other Ruislip woods.

CLIMATE CHANGE

Trees are one of the most abundant wild habitats in London, whether as sizeable woodlands, small copses, single specimens in parks and gardens, or avenues lining the road and railway routes. Providing nest and roost sites, foliage for shade, seasonal fruit, living homes for invertebrates or decaying wood for fungi and insects, trees support much of the capital's wildlife. Aesthetically they enrich our own lives, perhaps at their most uplifting when in full autumnal colour.

Yet more fundamentally, they support all life, for they are London's lungs and are vital for the health of the city. Not only do trees produce oxygen, providing air to breathe, but they absorb CO_2, responsible for global warming. With CO_2 emissions growing at an alarming rate, due to an increasingly energy-consuming world, we now face the reality of climate change and a potential environmental nightmare.

Global warming is already underway and temperatures are expected to rise faster in the south-east of the UK than further north. Winter rainfall is predicted to increase, but the chance of snow will decline. Summer rains will also decrease as the city's climate generally becomes drier. Global warming is expected to be noticeable within the next few decades. Birds and insects may adapt to the change by moving north to be replaced by species from the Continent. Black kites, egret and hoopoes may soon become regular sights in the capital. However, plants cannot move at the speed of climate change and some trees are expected to suffer seriously from the drier conditions. The beech, in particular, may disappear from London and much of the South-East.

Significantly reducing our energy and resource use and lowering emissions is the single biggest action for us all to take in combating global warming.

Supporting local conservation projects to protect valuable woodlands also helps trees neutralise carbon emissions. Exploring the wealth of wildlife on your doorstep rather than jetting off to the exotic is a very rewarding action, both personally and environmentally.

Two of London's most elegant birds show themselves briefly. The mottled brown treecreeper spirals up the trunk of an English oak. Supported by its tail, the slender creeper climbs up the bark, probing in crevices with its slender down-curved bill. Unable to climb down the trunk, once at the top, it flies downwards to the base of the next tree to be scaled.

Nearby, a loud 'twee-twee-twee' call announces the presence of another tree-climber. On the outer branches of an oak, a nuthatch scours the bark for insects. The stunning plumage of the nuthatch is instantly recognisable, blue grey on the upperparts, a black band across its eye and a pale chest, which becomes buff – almost orange – at its flanks. Unlike the treecreeper, the nuthatch can climb up and down trunks looking for food. It was originally known as 'nut-hack' because of the way it wedges acorns or nuts into cracks in bark, then hammers them open with its stout beak. Satisfied with its search for insects, the nuthatch flies briefly down to the ground, then back up to a decaying branch to continue its foraging.

The twenty-mile Hillingdon trail starts from just above Bayhurst Wood and continues south to Cranford near Heathrow. After Bayhurst, the footpath passes through the oak and hornbeam Mad Bess Wood, which some say is named after the wife an 18th-century gamekeeper. Supposedly the old woman prowled the woods at night looking for poachers.

The trail skirts through the edge of Copse Wood and alongside part of Ruislip Reservoir. With concrete sides, the reservoir is not overly attractive to waterside birds, but is home to many cormorants, geese and ducks. The grassy areas of Poor's Field are a good place to look for gatherings of small birds – passage migrants or flocking finches. There is even the chance of adders!

Oaks support more species of insects than any other native tree. The hundred-year-old oaks of Ten Acre Wood on the Hillingdon Trail are no exception and here a young tree plays host to a comma butterfly. Commas emerge late in the year and are still active in October and November. They survive by feeding on autumn flowering plants such as ivy.

Passing tussocks of soft rush and tufted hair grass, the Hillingdon Trail runs south from Ruislip Woods, through Ickenham Marsh – one of four reserves managed by the London Wildlife Trust in the Yeading Valley. The path continues past RAF Northolt Aerodrome and under the busy A40 dual carriageway towards Gutteridge Wood. Originally known as Great Hedge, Gutteridge is an area of ancient oak-hazel coppice, famed for its superb displays of bluebells in spring.

A short way further along the course of Yeading Brook is Ten Acre Wood, one of London's real gems. Approached by road, Ten Acre Wood lies hidden behind a maze of avenues and residential roads in Hillingdon – eluding all but the most inquisitive, or those in the know. A 19th century oak plantation with coppiced hazel below, the delightful woods and surrounding Cowslip Meadow are superb for wildlife watching. On the edge of Cowslip Meadow and Ten Acre Wood, this Midland hawthorn offers an abundance of scarlet

haws (berries). The haws of the hawthorn are a hardy berry and together with the hips of the wild dog rose will provide food for birds well into the winter.

In a good year, Ten Acre Wood is one of the best sites in London for owls; during late autumn this small habitat can boast four species of them. Tawny owls are resident, as is the little owl, which is best observed in its daytime roosts on some of the lone oaks standing in the meadow. Short-eared and long-eared owls are winter visitors to the valley; though short-eared occur more often than long-eared. However, in 1991-92, a bumper season, seven short-eared owls were present and over the same period, seventeen long-eared owls were counted. These were unusually high counts and numbers vary greatly from year to year. Late afternoons are the best time to look for owls, shortly before dusk. In autumn, the reserves are also good places to look for grey partridge, woodpecker, lapwing and seasonal migrants.

Across Charville Lane you come to the last of the four reserves, the tranquil Yeading Brook Meadows. Though surrounded by housing, the meadow holds an astonishing wealth of wildlife, offering a wonderfully accessible nature reserve to local communities. Surprisingly, snipe and skylark breed in the meadow, the latter being a species in serious national decline because of a change in agricultural practices. The plant life of the reserve is also remarkable. Long stretches of mature hawthorn and blackthorn screen the grasslands, their berries a great draw for winter birds and mammals.

At the far end of the meadow are two artificial hollows, dug after the war for use as a boating lake and children's paddling pool. Left unmaintained, they have since been colonised by water plants. In spring, the pools are adorned with brilliant yellow flag iris, spotted orchid, narrow-leaved dropwort and other flowers and grasses.

From the pools, the Hillingdon Trail continues further south, following the Grand Union Canal for a short way. Further on it meets the River Crane before reaching its final destination in Cranford Park.

Epping Forest is an epic forest, extending for 12 miles from Epping town to Wandstead Flats – the M25 motorway even runs underneath it. At almost 2,500 hectares it is the largest open space in the London area, though much of this falls outside the Greater London boundary. A former Royal hunting forest, the Queen Elizabeth Hunting Lodge, built for Henry VIII, can still be visited today. In the late 1800s, areas of the forest were being enclosed for development with little regard to the needs of local people. Concerned, the Corporation of London took action and in 1878 an Act of Parliament was passed, entrusting the ownership and care of Epping Forest to the Corporation. Today, millions of people visit the forest each year to enjoy the open space and wild habitats. Approximately two-thirds of the area is woodland – pollarded oaks, hornbeams and beeches are the dominant tree species.

"...the kestrel uses the old oak tree as a vantage point"

On a crisp autumnal morning, shortly after sunrise, a kestrel lands on a branch of a solitary oak tree. Surrounded by grassland, the kestrel uses the old tree as a vantage point to rest between hunting forays. With distinctive angular pointed wings it takes short flights, scouring the groundcover below looking for small mammals, birds and invertebrates. Once prey is located, the kestrel starts to hover, gradually dropping nearer to the ground in stages. As it hovers, its tail is pointed downwards for stability, which with precision wing beats allow the bird to keep its head perfectly still. After successive dips, it makes the final plunge downwards on to the prey. This characteristic hunting technique is most often witnessed along the roadside, particularly motorways, where it preys on small mammals living on these little disturbed wild margins. London's most abundant bird of prey, the kestrel is found in many of the capital's wild habitats. Wonderfully, it successfully breeds in the heart of the city, in St James's Park, Regent's Park and the grounds of 10 Downing Street. Away from the centre, good places to watch kestrels in London include Hampstead Heath, Richmond Park, the refuse tip at Rainham and Epping Forest.

Some of Epping's most unique wild residents are the dark fallow deer. Throughout the northern part of the forest, the wild roaming fallow deer vary in colour and markings, from the normal golden brown with white spots to the distinctive all-black animals. This black colouration of the fallow occurs naturally and it is thought that the first animals to be introduced to the forest were black. To preserve this coat colouration, a herd of black deer live in a special deer sanctuary near Theydon Bois.

Flies feast on the odorous cap of the appropriately named stinkhorn, one of the more bizarre species of fungi found in the capital. Hainault Forest is a fantastic location for a fungi foray, especially if the weather has been mild and damp. Well over one hundred species have been recorded in the forest, ranging from the plentiful but poisonous sulphur tuft to the pungent solitary stinkhorn. Emerging from a white egg in the leaf-litter, the stinkhorn grows to about 15 cm, its tip covered in dark foul-smelling spores. Easily smelt from some distance, flies are attracted to the distinctively shaped fungi to eat and disperse the spores.

Hainault Forest, an area of ancient woodland and a designated Site of Special Scientific Interest, straddles the border of Greater London and Essex. Historically the forest has a chequered past and was almost destroyed.

In the 12th century, Hainault was part of the much larger Forest of Essex, a hunting reserve created by Henry I. Commoners had the right to graze cattle and pigs in the forest, but hunting or disturbing deer would incur severe penalties. Predominantly wooded with hornbeam and oak, the forest was very valuable for its timber. Due to its incredibly hard wood, hornbeam had many uses including butchers' blocks and mallets and prior to steel being easily available, was even used for spokes and cogs. The origin of the tree's name is Anglo Saxon, or Old English – 'horn' means hard and 'beam' means tree. Hornbeam was also an important source of charcoal. The nearby area of Collier Row gets its name from the charcoal burners, which were known as colliers.

Unlike the coppiced trees in Ruislip and Highgate Woods, the hornbeams of Hainault and Epping Forest are pollarded – that is cut at the top of the trunk to produce a crown of new shoots above the ground. Pollarding is carried out in preference to coppicing, where it is necessary to prevent forest deer and domestic grazers from reaching and eating the fresh growth.

Oak was the prime timber used in the construction of buildings and boats. In the 13th century Henry III used more than four hundred trees from Hainault for building work at Westminster Palace. Oak was also used in the leather industry, its bark a good source of tannin used in preparing animal hides.

With the arrival of the Industrial Revolution came a reduced need for timber and in 1851, the Crown Lands Act was passed for the disafforestation of Hainault Forest. In just six weeks, 2,000 acres of woodland were destroyed and 100,000 trees felled. The cleared space became agricultural land and a number of farms were established. However, the public were outraged. Edward North Buxton launched a campaign to protect the remaining forest and in 1902, the London County Council bought the surviving woodland. A few years later, Hainault Forest was designated as a public open space.

Today, the forest is a beautiful woodland rich in flora and fauna, managed by the Woodland Trust and the London Borough of Redbridge. As well as hornbeam and oak, the forest also contains plenty of birch, beech and holly. Near the Lambourne End car park, a lone wild service tree can easily be viewed, its leaves turning a dark red in autumn. One of the rarest native trees in Britain, it is slow growing and until the last century was probably cut for use as charcoal. The service tree grows singly among other broad-leaved species and individual trees can be found in Hainault, Highgate and Epping Forest. Due to the way its bark peels off in rectangular strips, leaving a chequered effect, it is known as the chequer tree in Kent and Sussex. In these areas 'The Chequers' is a popular name for public houses.

Hainault's birdlife is equally notable – the forest is one of the key sites in London for the hawfinch, a little-observed finch that favours the fruit of the hornbeam. The forest is also a prime breeding location for marsh tit and other notable species including tawny owl, all three woodpecker species and even nightingale and wood warbler.

A four-legged newcomer to the woods is the muntjac deer. Originally from Asia, it has escaped from captivity and established itself as a wild species. The smallest of our deer species, the muntjac is a shy animal, perfectly at home in the thick undergrowth.

Only ten miles from the heart of London, Hainault Forest is a superb stretch of ancient forest. Now, almost one hundred years after the forest was saved from felling, the Woodland Trust is hoping to acquire adjoining farmland to extend it and provide a buffer. Replanted with native broadleaved trees and with some areas left open, the Woodland Trust hopes to restore this lost forest to its original glory as a mixed wood pasture.

An extract from the Ilford Recorder 1906

The addition of 800 acres of woodland (Hainault Forest)... by formal dedication on Saturday July 21st 1906, cannot fail to be of considerable interest to the people of Ilford. It is only necessary to think of what that extensive domain might have become had it been... covered with factories and sordid streets of unhappy factory toilers stretching up to our very boundaries. From such a disaster we are spared by the public spirit of Mr Edward North Buxton and those who have co-operated with him to preserve this ample beauty spot from the profanation of bricks and mortar. ... what a blessing it must be to these liberated toilers to escape for a brief spell from their mean and sordid environment into the freedom and fresh air and leafy beauty of the woods.

Scattered along the forest floor in Scadbury Park, decomposing tree stumps and wind-broken branches sustain fungi of many different shapes and sizes. The dappled woodland light intensifies their beauty. Many hundreds of fungi species occur in London; the best crops are yielded in autumn when the weather is mild and damp.

In south-east London, Scadbury's 120 hectares of pasture and mixed woodland allow for a great diversity of species. Hosting the fungi are ancient oaks, sweet chestnut, ash, alder, hazel, sycamore and birch.

"...these jewels of disorder and chaos"

Scaling a tree in search of insects, the secretive green woodpecker often hides on the blind side of the trunk. The largest and least arboreal of the three woodpeckers, it spends much time on the ground, probing for ants, beetles and grubs. Easily identified by its distinctive plumage colour, it can be quickly recognised in flight and by its call. Once commonly known as the 'yaffle' because of its loud laughing call, the green woodpecker has a low undulating flight, rising and falling with the beat of its wings.

The decaying remains of old trees are a rich resource for wildlife. They provide homes for numerous insects and, subsequently, food for many birds, like this starling. In a world where humans have an unhealthy obsession with clearing-up and tidying, it is vital that we save space for these jewels of 'disorder' and 'chaos'.

Trooping the Colour has its origins on the battlefield, though now it is a traditional London event where soldiers parade their colours (flags) to celebrate the Queen's official birthday. On the battlefields of Richmond Park, only a short distance from walkers, a stag thrashes bracken to adorn his antlers like flags. But unlike our summer ceremony, for deer, the autumnal battle is real. And it is not just the males that fight, females also clash to establish a hierarchy – the dominant doe will gain priority access to prime feeding spots.

Few areas of London are as atmospheric and audibly wild as Richmond Park early on an autumn morning. Home to about 300 of London's largest animal, the red deer, as well as 350 of its smaller relative, the fallow deer – the seasonal changes signal the start of the annual deer rut. As the warmth of summer subsides, mornings rapidly take on a real chill. At dawn, mist hangs low on the ground concealing the changing mood in the woods and grasslands. In top condition, the red deer stags begin to gather together a harem of hinds so that they may mate with as many females as possible. Hinds are only receptive for a 24 hour period. Young conceived now will be born in June/July next year.

To protect his assets from rival males, a stag must remain alert and fight off challengers by sound, sight and smell and, if all else fails, through physically battling his opponent. Stags defending harems must be vigilant 24 hours per day, leaving little time to eat. Males lose a lot of weight during this time and late on in the rut are most vulnerable to rivals.

Stags rely on smell to ward off competition – he who smells strongest wins. Throughout Richmond Park are small pools that the males use for wallows. Urinating on the mud, they coat themselves with the pungent mix and with a favourable wind, can be smelt from quite a distance.

Only when stags are evenly matched will fighting occur. Still vocalising, the males move parallel to each other, then turn inwards and clash their antlers. Now a battle of strength, they lock antlers trying to force the other backwards. Very quickly the weaker stag is defeated and chased away.

"...mist allows minimal visibility, but the sounds are incredible"

Standing in the grasslands at sunrise on a frosty October morning, the mist allows minimal visibility, but the sounds are incredible. From three directions, red deer stags roar and grunt, announcing their location. Exaggerated by the inability to see, the intensity and anonymity of the strange sounds conjures up prehistoric images.

As the sun rises and the mist starts to lift, the clarity of the competition unfolds. Standing near to his herded females, a bellowing stag proclaims his dominance. The stag's sounds fill the grassland – his breath visible in the cold air. Across the grassland are two rival males strutting their stuff. This time the stag successfully holds his ground, but it will not be the last time he is challenged.

Autumn leaves with a bang, the last splash of colour explodes in the evening sky during the Bonfire Night celebrations. With shortened days and cold weather, bird activity begins to change. For residents such as the rook, winter is a time to form large feeding flocks, gathering to roost at sunset on a row of ancient oaks. Migrant species such as redwing and

fieldfare have already arrived from their northern breeding grounds and are busily devouring the hedgerow berries. Throughout the capital, thousands of waterfowl and waders are descending on the city's wetlands.

And so the wild cycle of life in London continues...

17
19
6
1
18
15
7
12
2
16
3
13
14
10
34
31
29
9
8
11
40
30
38
42
32
43
4
35
41
36
33
5
39
28
37

Wild Guide to London

Take a walk on London's wild side. London is rich with accessible wild places where people can enjoy the nature of the capital. The diversity of habitats in the city is incredible. Explore medieval marshes, constructed wetlands and urban creeks. Absorb the beauty and tranquillity of ancient woodlands, suburban heaths and chalk downland. Or marvel at the adaptability of wildlife – frogs in the West End, families of foxes in cemeteries and falcons nesting on tower blocks.

Enjoy a backyard safari

Here are a small selection of the nature reserves and green spaces mentioned in this book

Almost all of the sites are easily reached using public transport and the nearest rail/tube station is listed. However it may be easier to access particular sites by bus, so contact local operators for the best route. Access hours and admission charges are subject to change – it is advisable to check before visiting.

Neither the Publisher nor Author can accept liability for changes to the information given.

* **SSSI** – *Site of Special Scientific Interest*

■ LONDON – North of the Thames

1. Ruislip Woods National Nature Reserve SSSI *

London's first National Nature Reserve incorporates the oak and hornbeam dominated Bayhurst, Mad Bess, Copse and Park Woods and Poor's Field.

Further Details: www.english-nature.org.uk
Access: Open at all times. Free admission
Nearest rail/tube station: Ruislip tube, then bus

2. Denham Lock Woods SSSI *

Wet woodland along the Grand Union Canal. Part of a larger wild space that includes Denham Country Park, Frays Farm Meadows SSSI and the Colne and Fray rivers.

Further Details: www.wildlondon.org.uk
Access: Open at all times. Free admission
Nearest rail/tube station: Denham rail or Uxbridge tube, then bus along A40, then walk

3. Yeading Valley

Comprises four nature reserves managed by the London Wildlife Trust. The oak woodlands of Ten Acre Woods, oak-hazel coppice in Gutteridge Wood, Ickenham Marsh and the flower-rich Yeading Brook Meadows.

Further Details: www.wildlondon.org.uk
Access: Open at all times. Free admission
Nearest rail/tube station: Best reached by foot along the Hillingdon Trail. Alternatively, Ickenham tube, then walk along Glebe Avenue; Hillingdon tube, then bus

4. Crane Park Island

Formerly the Hounslow Gunpowder Mills, the island nature reserve and surrounding River Crane are now home to the increasingly scarce water vole.

Further Details: www.wildlondon.org.uk
Access: Open at all times. Free admission
Nearest rail/tube station: Whitton rail then walk along River Crane

5. Bushy Park

North of Hampton Court, this Royal park is home to herds of fallow and red deer, also many birds and amphibians.

Further Details: www.royalparks.gov.uk
0208 979 1586
Access: 6.30 am – Dusk.
Free admission
Nearest rail/tube station: Teddington, Hampton Wick, Hampton Court rail

6. Totteridge Fields

A large area of ancient meadow – the floral displays are fantastic in spring and summer.

Further Details: www.wildlondon.org.uk
Access: Open at all times. Entrance on Hendon Wood Lane.
Free admission
Nearest rail/tube station: Totteridge and Whetstone tube, then bus along Barnet Rd.

7. Welsh Harp Reservoir SSSI *

Also known as Brent Reservoir, the large expanse of open water, marshes, trees and grassland offer rewarding birdwatching, especially during the winter months.

Further Details: www.brentres.com
Access: Open at all times.
Enter from Cool Oak Lane.
Free admission
Nearest rail/tube station: Hendon rail

8. Gunnersbury Triangle

Cut off from the surrounding area by railway tracks, this small, but magical, wild haven was one of the London Wildlife Trust's first reserves.

Further Details: www.wildlondon.org.uk
Access: Open at all times. Entrance on Bollo Lane. Free admission
Nearest rail/tube station: Chiswick Park

9. Holland Park

Tranquil woodlands, grassy glades and ornamental gardens

Further Details: www.rbkc.gov.uk
Access: 8 am – Dusk.
Free admission
Nearest rail/tube station: Holland Park tube

10. Royal Parks – St James's Park, Green Park, Hyde Park and Kensington Gardens

Cutting through the centre of London, the four Royal parks form a massive green space for people and wildlife.

Further Details: www.royalparks.gov.uk

Access: St James's, Green and Hyde Park 5 am – Midnight, Kensington Gardens 6 am – Dusk. Free admission to all parks

Nearest rail/tube station: St James's Park, Green Park, Hyde Park Corner, Lancaster Gate and Bayswater tube

11. Natural History Museum Wildlife Garden

In a corner of the museum grounds, eight distinct habitats form this gorgeous wild space.

Further Details: www.nhm.ac.uk
0207 942 5011

Access: Summer, Noon- 5 pm. Free admission

Nearest rail/tube station: South Kensington and Gloucester Road tube

12. Hampstead Heath SSSI *

Only a few miles from Central London, the Heath boasts an impressive habitat range and diversity of flora and fauna.

Further Details: www.cityoflondon.gov.uk

Access: Open at all times except Golders Hill. Free admission

Nearest rail/tube station: Hampstead Heath and Gospel Oak rail

13. Regent's Park

The park's grasslands and lake allow some incredibly varied and close wildlife viewing – particularly the heronry.

Further Details: www.royalparks.gov.uk.

Access: 5 am – Dusk. Free admission

Nearest rail/tube station: Baker Street, Regent's Park and Camden Town tube

14. Phoenix Gardens

A small community garden in the heart of the West End – just off Charing Cross Road.

Further Details: www.thephoenixgarden.ik.com
happy.gardener@btinternet.com

Access: 8.30 am – Dusk. Entrance on Stacey Street. Free admission

Nearest rail/tube station: Tottenham Court Road or Leicester Square tube.

15. Highgate Wood

Oak and hornbeam woodlands that support a profusion of insects, bats, birds and fungi. Information Centre.

Further Details: www.cityoflondon.gov.uk

Access: 7.30 am – Dusk. Free admission

Nearest rail/tube station: Highgate tube

16. Camley Street Natural Park

Two unique wild acres of woodland and pond in the heart of King's Cross. Information Centre.

Further Details: www.wildlondon.org.uk
0207 833 2311

Access: Mon – Thurs 9-5, Fri closed, Sat/Sun 11-5. Free admission

Nearest rail/tube station: King's Cross and St Pancras rail or tube

17. Lee Valley Park

The park stretches from the River Thames at East India, to Ware in Hertfordshire. The large area provides a great diversity of specialities – from orchids to waterbirds.

Further Details: www.leevalley.org.uk

Access: Open at all times.

Nearest rail/tube station: Many along route.

18. Walthamstow Reservoirs SSSI *

A large group of reservoirs owned by Thames Water, many are accessible by the public. Heronries, cormorant colony and excellent water bird watching.

Further Details: www.thameswater.co.uk
0845 9200 800

Access: Daytime. Small admission charge.

Nearest rail/tube station: Tottenham Hale tube or Blackhorse Road rail.

19. Epping Forest SSSI *

Though mostly outside Greater London, the forest is one of the area's most important wild sites. The ancient woodland and grassy areas support an incredible array of species. Information centre.

Further Details: www.cityoflondon.gov.uk

Access: Open at all times. Free admission

Nearest rail/tube station: Chingford rail

20. Hainault Forest

Ancient forest, mainly hornbeam and oak, good for birds and fungi.

Further Details: www.woodland-trust.org.uk

Access: Open at all times. Free admission

Nearest rail/tube station: Limited accessibility from Hainault or Grange Hill tube

21. The Chase

The Chase is London Wildlife Trust's largest nature reserve. Habitats include shallow wetlands, reedbeds, pasture, scrub and woodland – home to the rare black poplar tree.

Further Details: www.wildlondon.org.uk
0208 593 8096

Access: Open at all times. Free admission

Nearest rail/tube station: Dagenham East tube then bus along Dagenham Rd.

22. The Ripple

Once the dumping ground for pulverised fuel ash, the site is now a superb place for orchids and other wild flowers.

Further Details: www.wildlondon.org.uk

Access: Open at all times. Entrance on Thames Rd/Renwick Rd. Free admission

Nearest rail/tube station: Barking tube or rail then bus to Thames View Estate

23. Duck Wood

Impressive displays of bluebells in spring and the chance of badger, fox and fallow deer year round.

Further Details: www.havering.gov.uk

Access: Open at all times. Free admission

Nearest rail/tube station: Harold Wood rail then bus

24. Rainham Marshes Nature Reserve SSSI *

An RSPB reserve, the superb marshland habitat along the edge of the Thames is visited by thousands of water birds and is home to good water vole populations.

Further Details: www.rspb.org.uk/rainham
01708 892900

Access: Open in summer 2006. Prior to this, special advance events can be booked. Admission charge

Nearest rail/tube station: Purfleet rail, then walk

LONDON – The Thames

25. The River

Don't forget the Thames – from Hampton Court to Rainham Marshes the river hosts some amazing wildlife. View from its banks or by boat.

Further Details: www.environment-agency.gov.uk

Access: At all times along the Thames Path. Free admission

Nearest rail/tube station: Many along route

LONDON – South of the Thames

26. Oxleas Wood SSSI *

A large expanse of ancient woodland and meadows famously saved from development.

Further Details: www.english-nature.org.uk

Access: At all times. Free admission

Nearest rail/tube station: Falconwood rail

27. Scadbury Park

Damp woodlands and grassy glades supporting a profusion of wildlife.

Further Details: www.bromley.gov.uk

Access: Open at all times. Free admission

Nearest rail/tube station: Chiselhurst or Sidcup rail, then bus along Perry Street

28. Saltbox Hill SSSI *

A beautiful area of chalk downland and one of the richest sites in Greater London for wildflowers (especially orchids), grasses and butterflies.

Further Details: www.wildlondon.org.uk

Access: Open at all times. Free admission. Entrance on Hanbury Drive, Westerham Rd.

Nearest rail/tube station: Bromley North rail then bus to Biggin Hill Airport

29. Greenwich Ecology Park

Two lakes surrounded by a range of habitats, from shingle beach to marsh and wet woodland. Information centre.

Further Details: www.urbanecology.org.uk

Access: Wednesday- Sunday 10 am – 5 pm. Free admission

Nearest rail/tube station: North Greenwich tube

30. The Creekside Centre

Exploring Deptford Creek is a unique wild experience – the tidal movements create an exciting wild habitat.

Further Details: www.creeksidecentre.org.uk
0208 692 9922

Access: Booking in advance essential.

Nearest rail/tube station: Greenwich DLR and rail

31. Stave Hill Ecology Park and Lavender Pond Nature Park

Two superb urban reserves created on the sites of derelict dockland. Ponds, meadows, wetlands, woodlands and scrub.

Further Details: www.urbanecology.org.uk

Access: Open at all times. Free admission

Nearest rail/tube station: Canada Water or Surrey Quays tube

32. Nunhead Cemetery Nature Reserve

This Victorian Cemetery is a wild oasis in an urban area - well known for its foxes, birds and wild flowers.

Further Details: www.fonc.org.uk
Access: 8.30am - 4.30pm. Free admission
Nearest rail/tube station: Nunhead rail

33. Beckenham Place Park

A prime wild space, comprising many habitats - ancient woodland, acid grasslands and riverbank.

Further Details: www.lewisham.gov.uk
Access: Open at all times. Free admission
Nearest rail/tube station: Ravensbourne or Beckenham Hill rail

34. Tate Modern (Outside)

From the grounds look to the top of the power station chimney for roosting peregrines, or out across the Thames for birds and even seals.

Further Details: 0207 887 5120
Access: Views of the chimney can be seen from the Thames footpath at all times. Free admission
Nearest rail/tube station: London Bridge rail or tube, Southwark tube

35. Sydenham Hill Wood

A wonderful mix of ancient and recent woodland - home to over 200 tree species and an equally impressive array of birds, mammals, fungi and flowers.

Further Details: www.wildlondon.org.uk
0208 699 5698
Access: Open at all times. Free admission
Nearest rail/tube station: Sydenham Hill rail (itself managed for wildlife)

36. Dulwich Upper Wood

A fragment of the old Great North Wood - great for birds, fungi and bluebells.

Further Details: www.urbanecology.org.uk
Access: Open at all times. Free admission. Entrance on Farquhar Rd
Nearest rail/tube station: Gipsy Hill or Sydenham Hill rail

37. Coulsdon Coppice

Peaceful woodland surrounded by a residential area. Wildflowers and insects thrive in the central grassy glades.

Further Details: www.wildlondon.org.uk
Access: Entrance off St David's. Open at all times. Free admission
Nearest rail/tube station: Coulsdon South rail

38. Battersea Park

Areas of the park are specifically managed for wildlife - a great place for butterflies and birds, particularly cormorants by the Thames.

Further Details: www.batterseapark.org
Access: Open at all times. Free admission
Nearest rail/tube station: Battersea Park or Queenstown Road rail

39. Morden Hall Park

A large area of parkland managed by the National Trust. The hay meadows, wetlands and banks of the River Wandle are especially beautiful in spring.

Further Details: www.nationaltrust.org.uk
Access: 8 am -6 pm. Free admission
Nearest rail/tube station: Morden tube

40. London Wetland Centre, Barnes SSSI *

The Wildfowl and Wetlands Trust's award-winning reserve offers more than 40 hectares of created wetlands in the heart of the city. Information centre.

Further Details: www.wwt.org.uk
Access: 9.30 am - 5 pm, open Thursday evenings in summer. Admission charge
Nearest rail/tube station: Hammersmith tube or Barnes rail then bus or walk

41. Wimbledon Common SSSI *

The Commons consist of woodland, scrubland, heathland and nine ponds.

Further Details: www.wpcc.org.uk
020 8788 7655
Access: Open at all times. Free admission
Nearest rail/tube station: Wimbledon tube or rail, then bus.

42. Royal Botanic Gardens, Kew

Areas of the gardens are managed for wildlife and attract an impressive variety of butterflies, dragonflies and birds

Further Details: www.rbgkew.org.uk
0208 332 5655
Access: Opens at 9.30 am, closing times vary. Admission charge
Nearest rail/tube station: Kew Gardens rail and tube

43. Richmond Park National Nature Reserve SSSI *

The largest open space in London - superb for a diversity in any season and home to a thousand species of beetle.

Further Details: www.royalparks.gov.uk
0208 948 3209
Access: 7.30 am - Dusk. Free admission
Nearest rail/tube station: Richmond rail then bus to Richmond Park Gate

Organisations working for wildlife in London

English Nature

www.english-nature.org.uk
Northminster House,
Peterborough PE1 1UA

London Wildlife Trust

www.wildlondon.org.uk
Skyline House, 200 Union Street,
London, SE1 0LW
0207 261 0447

London Natural History Society

www.lnhs.org.uk

Royal Parks

www.royalparks.gov.uk
The Old Police House, Hyde Park,
London W2 2UH
0207 298 2000

The Royal Society for the Protection of Birds (South-East)

www.rspb.org.uk
South East Regional Office, 2nd Floor,
Frederick House, 42 Frederick Place,
Brighton BN1 4EA
01273 775333

Trust for Urban Ecology (TRUE)

www.urbanecology.org.uk
The Ecology Park, Thames Path,
John Harrison Way, Greenwich Peninsula,
London SE10 0QZ
0208 293 1904

Wildfowl and Wetlands Trust (London)

www.wwt.org.uk
London Wetland Centre,
Queen Elizabeth's Walk, Barnes,
London SW13 9WT
0208 409 4400

Woodland Trust

www.woodland-trust.org.uk
Autumn Park, Grantham,
Lincolnshire NG31 6LL
01476 581111

Enjoy and let others enjoy London's wild spaces

- Avoid disturbing wildlife
- Keep to public paths
- Keep dogs under control
- Leave nothing but footprints
- Take nothing but photographs